# HOW THE
# UNIVERSE
# WORKS

## AN ILLUSTRATED GUIDE TO THE COSMOS AND ALL WE KNOW ABOUT IT

# HOW THE UNIVERSE WORKS

## AN ILLUSTRATED GUIDE TO THE COSMOS AND ALL WE KNOW ABOUT IT

CHARTWELL
BOOKS

Brimming with creative inspiration, how-to projects, and useful information to enrich your everyday life, Quarto Knows is a favorite destination for those pursuing their interests and passions. Visit our site and dig deeper with our books into your area of interest: Quarto Creates, Quarto Cooks, Quarto Homes, Quarto Lives, Quarto Drives, Quarto Explores, Quarto Gifts, or Quarto Kids.

**Original Idea** Nuria Cicero
**Editorial Coordination** Alberto Hernández
**Design** Caco Daví, Leandro Jema
**Editing** Joan Soriano
**Layout** Clara Miralles, Laura Ocampo

Printed in China
10 9 8 7 6 5 4 3 2 1

MIX
Paper from
responsible sources
FSC® C101537

# PREFACE

Featuring amazing photography and stunning full-color illustrations, *How the Universe Works* will take you on a journey into the Cosmos and reveal all its mysteries. Hundreds of subjects are explored, thousands of facts are revealed, explained, and dissected so that anyone can understand concepts as simple as why the Moon causes tides or as complex as the Big Bang. Other complicated concepts—including the mystery of dark matter, the general theory of relativity, how black holes are created, and if time travel might ever be possible— are also clearly described.

Visually-driven and encyclopedic in scope, easy to understand for all readers and even of interest to experts, *How the Universe Works* is organized thematically into eight chapters, and includes topics such as the physical nature of the Cosmos, the composition of the Solar System, the Earth and Moon, the history of astronomy, the Space Race, exploration of other planets, and even technological concepts related to space. Utilizing scientific imagery and materials courtesy of the world's space agencies, you will spend many enjoyable hours learning about the nature of gravity, the Sun, the history of the Apollo program, the effects of space travel on the human body, and the planned missions to Mars, among other fascinating topics. So come explore the amazing facts and undiscovered secrets of our vast and limitless Cosmos.

# CONTENTS

INTRODUCTION                                9

## CHAPTER 1
## THE SECRETS OF THE UNIVERSE

Radiography of the Cosmos                   16
The Moment of Creation                      18
Forces of the Universe                      22
Theories About the Future                   24
Anatomy of the Galaxies                     26
Our Galaxy: The Milky Way                   28
Active Galaxies                             30
With Their Own Light                        32
Stellar Evolution                           34
Red Giants                                  36
Gas Shells                                  38
Supernova                                   40
The Final Darkness                          42

## CHAPTER 2
## THE SOLAR SYSTEM

The Solar System                            46
A Very Warm Heart                           48
Mercury                                     50
Venus                                       52
Mars                                        54
Jupiter                                     56
Saturn                                      58
Uranus                                      60
Neptune                                     62
Pluto                                       64
Distant worlds                              66
Asteroids and Meteorites                    68
Comets                                      70

Extrasolar Planets                          72
Like the Earth                              74

## CHAPTER 3
## THE EARTH AND THE MOON

The Blue Planet                             78
Journey to the Center of the Earth          80
The Earth, a Huge Magnet                    82
Surface and Movements of the Moon           84
The Moon and The Tides                      86
Eclipses                                    88
Traversing Time                             90
Under Construction                          94
Scorching Flow                              96
Minerals: "The Bricks"                      98
The Origin of Life                          100
Fossil Relics                               102
Cambrian Explosion                          104
Conquest of the Earth                       106
The Tree of Live                            108
Other Origin?                               110

## CHAPTER 4
## HISTORY OF ASTRONOMY

The First Astronomers                       114
Astronomical Theories                       116
Sprinkled with Stars                        118
The Astronomical Clock of Su Song           120
Copernicus and Galileo                      122
Newton and the Universal Gravitation        124
The Impact of Relativity                    126
Large Hadron Collider                       128
Space Observatory                           130
Celestial Cartography                       132
From the Back Yard                          134

## CHAPTER 5
## THE SPACE RACE

| | |
|---|---|
| From Fiction to Reality | 138 |
| NASA | 140 |
| Control from Earth | 142 |
| Other Space Agencies | 144 |
| Russian Missions | 146 |
| United States Spacecraft | 148 |
| One Giant Leap for Mankind | 150 |
| The Apollo Program | 152 |
| Echoes of the Past | 154 |

## CHAPTER 6
## EXPLORATION MISSIONS

| | |
|---|---|
| Interest in the Planets | 158 |
| Point of Departure | 160 |
| Rockets | 162 |
| Takeoff Chronicle | 164 |
| Space Shuttle | 166 |
| Profession: Astronaut | 168 |
| Far from Home | 170 |
| Space Stations | 172 |
| Spying on the Universe | 174 |
| The Chandra Observatory | 176 |
| Voyager Probes | 178 |
| SETI Projects | 180 |
| Closer to the Sun | 182 |

## CHAPTER 7
## TRAVELING TO MARS AND OTHER WORLDS

| | |
|---|---|
| Mars Missions | 186 |
| Mars in the Sights | 188 |
| Mars Reconnaissance Orbiter | 190 |
| Martian Rovers | 192 |
| Curiosity | 194 |

| | |
|---|---|
| Is it Possible to Colonize Mars | 196 |
| A Film Challenge | 198 |
| Jupiter in Focus | 200 |
| The Moon Europa | 202 |
| A View of Saturn | 204 |
| Toward Venus and Pluto | 206 |

## CHAPTER 8
## CONNECTED WITH SPACE

| | |
|---|---|
| Space Technology at Home | 210 |
| Microgravity and Science | 212 |
| Global Interconnection | 214 |
| Satellite Orbits | 216 |
| Environmental Satellites | 218 |
| Earth Images in High Resolution | 220 |
| Space Junk | 222 |
| Global Satellite Navigation | 224 |
| Space Vacations | 226 |
| Stratospheric Adventure | 228 |
| A Colossal Performance | 230 |

## INDEX

| | |
|---|---|
| Index | 233 |
| Photo credits | 240 |

**CRAB NEBULA**
Located more than 6,000 light years from Earth, this nebula emerged from the explosion of a star that occurred about 900 years ago.

# INTRODUCTION

Since the beginning of time, humans have been curious about the celestial unknown. That powerful desire to know led us to believe that the stars were tribal campfires in the night sky, that the universe was flat and resided on top of a giant tortoise shell, and that the Earth, as described by the Greek astronomer Ptolemy, was at the center of the universe. More recently, curiosity has propelled the development of sophisticated telescopes and spacecraft that allow us to peer with even greater clarity at blurry and distant objects and to trek farther into the unknown. *How the Universe Works* will take you along on that journey into the Cosmos, illustrated with spectacular photographs that unfold the nature of the planets and stars adorning the night sky. You will learn how Sun-like stars are formed and die, discover the nature and design of black holes, encounter the mysterious dark matter that surrounds the galaxies, and come to understand our place in the immensity of the Universe we occupy. By comparing our own evolution with that of other worlds, you will fully understand why, for now at least, there is no better place for humans to live than Earth.

It is estimated that our Sun is one of 400 billion stars in the Milky Way galaxy. Given such immensity, could it be that our solar system is the only one with a habitable planet? Every day astronomers grow more certain of the possible existence of other worlds capable of hosting life. They only need to find them. In recent years, huge steps have been taken in locating almost 3,500 exoplanets and

① 
## SOLAR SYSTEM
**Scientists believe its origin may be around 4.6 billion years ago.**

② 
## TARANTULA
**Emission nebulae, like this one, are clouds of hot gas and dust that glow in space.**

③ 
## SPACE RACE
**In the 1950s the United States and the Soviet Union began the era of space exploration.**

numerous solar systems similar to ours, such as TRAPPIST-1, an ultra-cool dwarf star with seven temperate planets located 39.5 light years from our Sun. Although space observation goes back to ancient times, space exploration is relatively new. The first satellite was launched into space in 1957, and within just a little over a decade astronomers observed other frozen worlds, much smaller than planets, in a region called the Kuiper belt. Scientists assure us that we live in interesting times when it comes to the exploration of our solar system, especially considering all that has been discovered in just the past 50 years, and estimate that even more discoveries will come at an accelerated rate.

*How the Universe Works* will take you on a detailed trip into the mysteries of our solar system. Mars has piqued specific scientific interest due to its proximity to Earth and the probability that it once sustained life. The space orbiters Mars Odyssey and Mars Express have confirmed

there are ice deposits within the greater depths of the red planet. Sending exploratory spacecraft to Saturn has been another significant scientific achievement, demonstrating our immense capacity to dream of new worlds. Meanwhile, the New Horizons mission is exploring the outer edges of our solar system by venturing on a reconnaissance mission to the dwarf planet Pluto. Another challenge is to successfully colonize other planets, with Mars as the short-term key objective. Assuming the technical and economic obstacles can be overcome, there are various private projects like Mars One interested in establishing the first human settlement off of Earth. But *How the Universe Works* has just begun, there is still so much more to discover.

Naturally, we will pause to carefully analyze Earth, to better understand its origin and formation, its evolution, its characteristics,

④
### MOON
**The Earth's only permanent natural satellite is located 238,900 mi away.**

②

①

## CHANDRA

**The space observatories have made it possible to analyze the universe with greater precision.**

and its relation to the Sun and Moon. We will study our neighbors, the other planets in our solar system, their satellites and the important characteristics that distinguish them. You will also learn about meteorites, asteroids and comets that orbit around the Sun. All this information—a microscopic exploration of the mysteries of the Universe—is accompanied by groundbreaking scientific imagery courtesy of the principal space agencies. These images, captured by the latest space telescopes, allow us to see and better understand the surface of each extraterrestrial object and their volcanoes and craters. All of the imagery and illustrations accompanying the text—like, for example, star maps in which we see constellations, the star clusters that since ancient

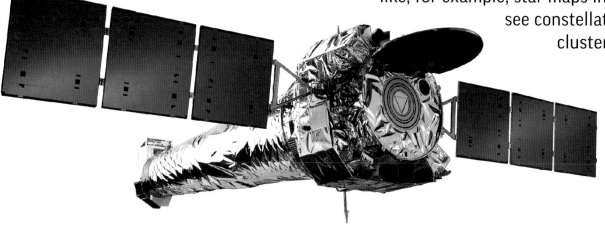

times have guided humans in navigation and the creation of calendars—are a great help when studying and understanding the structures of objects both visible and invisible (such as dark matter) that are integral foundations of the Cosmos.

Traveling back in time, we will review the history of space starting with Ptolemy, who believed that planets orbited the Earth, then Copernicus, who placed the sun at the center of the universe, and on to Galileo, who was first in reaching to the skies with a telescope, and the most recent theories of Stephen Hawking, the genius of space and time who continues to astound us with his discoveries of humankind's greatest scientific mysteries. We will also review all that space exploration has given to our everyday lives on Earth, including the utilization of satellites, which have revolutionized everything from the use of cell phones, the Internet and televisions to GPS navigation. So come along and discover *How the Universe Works.*

② 
**SPACE STATIONS**
Films like *2001: A Space Odyssey* anticipated this type of habitable space module.

③ 
**CONSTELLATIONS**
Astronomers have cataloged 88 constellations throughout history.

④ 
**CHALLENGES**
Reaching Earth's orbit was not simple: it required the dedication of scientists and the courage of astronauts.

**NGC 1300**
This barred spiral galaxy is 61 million light years from Earth and has a size similar to the Milky Way.

## CHAPTER 1

# THE SECRETS
# OF THE UNIVERSE

Scientists estimate that there are at least two trillion galaxies in the universe. The Milky Way—the galaxy where the Earth is located—is like a drop of water in a vast ocean.

# RADIOGRAPHY OF THE COSMOS

**The Universe, or cosmos, which marvels us with its majesty, is a series of one hundred billion galaxies.** In turn, each of these galaxies, which tend to join together in large groups, contains billions of stars. These galactic concentrations are surrounded by empty spaces, or 'cosmic gaps.'

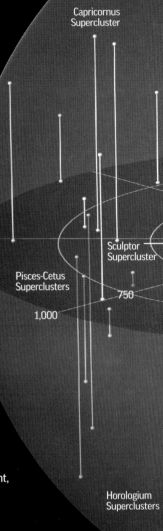

Capricornus Supercluster

Sculptor Supercluster

Pisces-Cetus Superclusters

750

1,000

Horologium Superclusters

## ① EARTH

**Created along with the Solar System when the Universe was already nine billion years old.**

## THE UNIVERSE

Dating back almost 14 billion years to a gigantic explosion, it is impossible to render an idea of the current size of the Universe. And the infinite number of stars and galaxies remaining continues to expand. For many years, astronomers believed the Milky Way represented the entire Universe. However, during the twentieth century, it was discovered that not only is space much more vast than originally thought, but that it was located within an expanse of extraordinary dimensions.

## ② NEARBY STARS

**Located 20 light years from the Sun in all directions, they form the 'solar neighborhood.'**

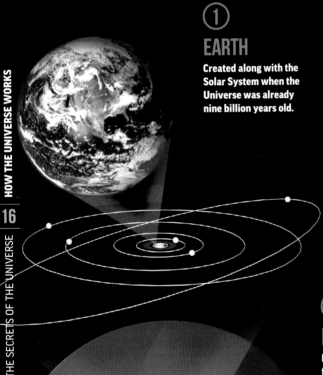

G51-15
Ross 128
Lalande 2185
Wolf 359
Struve 2398
Procyon
12.5
7.5
90°
Barnard's Star
Luyten's Star
2.5
**SUN**
Alpha Centauri
61 Cygni
Sirius
Groombridge 34
Ross 248
0°
270°
Ross 154
Epsilon Eridani
L726-8
Epsilon Indi
L789-6
L372-58
Tau Ceti
L725-32
Lacaille 9352

## ③ NEIGHBORS

**The Milky Way and its nearest galaxies are within one billion light years.**

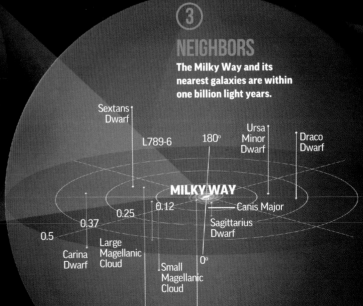

Sextans Dwarf
L789-6
180°
Ursa Minor Dwarf
Draco Dwarf
**MILKY WAY**
0.12
Canis Major
0.25
Sagittarius Dwarf
0.37
0°
0.5
Carina Dwarf
Large Magellanic Cloud
Small Magellanic Cloud

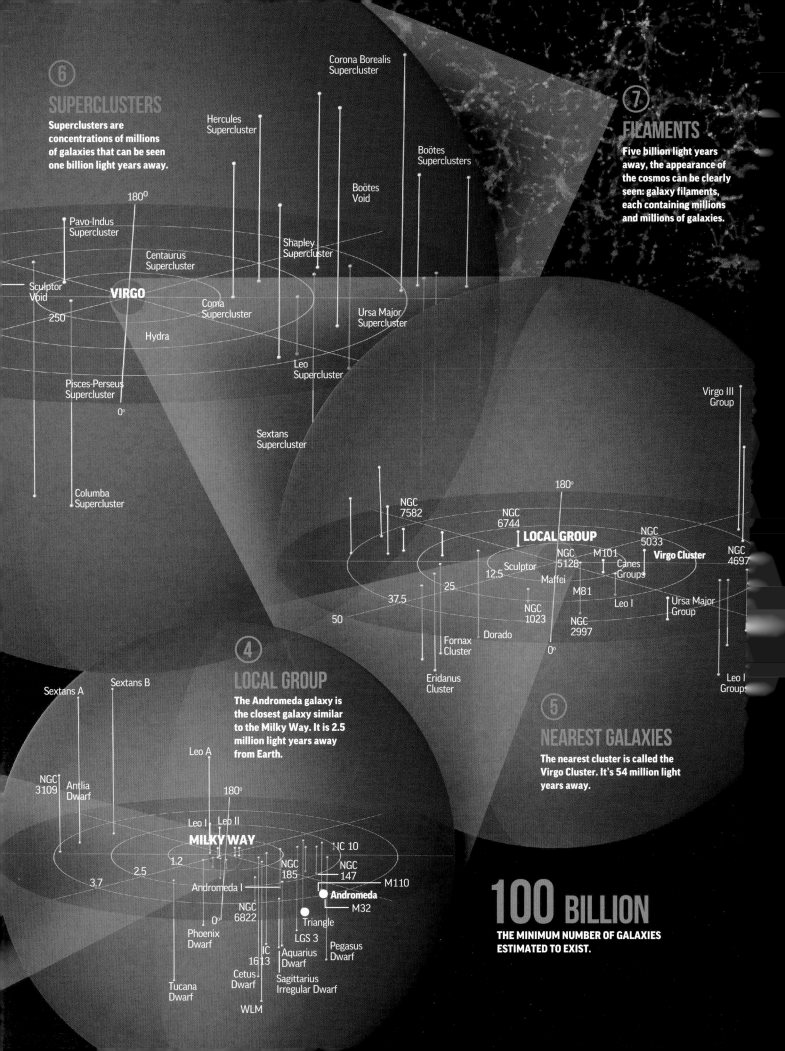

**6**

**SUPERCLUSTERS**

Superclusters are concentrations of millions of galaxies that can be seen one billion light years away.

**7**

**FILAMENTS**

Five billion light years away, the appearance of the cosmos can be clearly seen: galaxy filaments, each containing millions and millions of galaxies.

Pavo-Indus
Supercluster

180°

Hercules
Supercluster

Corona Borealis
Supercluster

Boötes
Superclusters

Centaurus
Supercluster

Sculptor
Void

**VIRGO**

250

Shapley
Supercluster

Boötes
Void

Coma
Supercluster

Hydra

Ursa Major
Supercluster

0°

Leo
Supercluster

Pisces-Perseus
Supercluster

Sextans
Supercluster

Virgo III
Group

Columba
Supercluster

NGC
7582

NGC
6744

180°

**LOCAL GROUP**

NGC
5033

NGC
4697

NGC
5128

M101

Canes
Groups

**Virgo Cluster**

Sculptor

12.5

Maffei

37.5

25

M81

Leo I

Ursa Major
Group

50

NGC
1023

Dorado

NGC
2997

0°

Leo I
Groups

Sextans A

Sextans B

**4**

**LOCAL GROUP**

The Andromeda galaxy is the closest galaxy similar to the Milky Way. It is 2.5 million light years away from Earth.

Fornax
Cluster

Eridanus
Cluster

**5**

**NEAREST GALAXIES**

The nearest cluster is called the Virgo Cluster. It's 54 million light years away.

NGC
3109

Antlia
Dwarf

Leo A

180°

Leo I Leo II

**MILKY WAY**

1.2

IC 10

2.5

NGC
185

NGC
147

3.7

Andromeda I

M110

**Andromeda**

M32

0°

NGC
6822

Triangle

Phoenix
Dwarf

IC
1613

LGS 3

Pegasus
Dwarf

Tucana
Dwarf

Cetus
Dwarf

Aquarius
Dwarf

Sagittarius
Irregular Dwarf

WLM

**100 BILLION**

**THE MINIMUM NUMBER OF GALAXIES ESTIMATED TO EXIST.**

# THE MOMENT OF CREATION

**It is impossible to know with any level of accuracy how, from nothing, the Universe was born.** Initially, according to the Big Bang theory – the most commonly accepted theory among the scientific community – an infinitely small and dense hot ball appeared, which gave rise to space, matter and energy. This happened 13.7 billion years ago, although what generated it is unknown to this day.

## HOW IT GREW
The inflation caused each region of the young Universe to grow. The galactic neighborhood appears uniform: the same types of galaxy, the same background temperature.

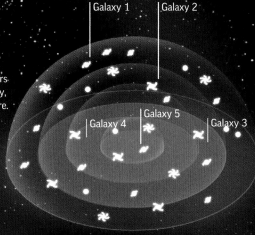

Galaxy 1 Galaxy 2 Galaxy 5 Galaxy 4 Galaxy 3

## ENERGETIC RADIATION
The hot ball that gave rise to the Universe was a permanent source of radiation. Subatomic particles and antiparticles annihilated one another. The high density spontaneously created and destroyed matter. Having remained in this state, the Universe would never have experienced the growth that, as is believed, occurred as a result of 'cosmic inflation.'

| TIME | 0 | $10^{-43}$ SECONDS | $10^{-38}$ SECONDS |
|---|---|---|---|
| **TEMPERATURE** | ① - | ② $10^{32°}$ C | ③ $10^{29°}$ C |

① According to the theory, everything that currently exists was compressed into a space smaller than the nucleus of an atom.

② At the moment closest to hour 'zero,' which physics has been able to identify, the temperature is immensely high. A superforce governed the Universe.

③ The Universe is unstable and grows 100 trillion trillion trillion trillion trillion times. Inflation starts and the forces separate.

## ELEMENTARY PARTICLES
Initially, the Universe was a 'hodgepodge' of particles that interacted with others due to high levels of radiation. Later, once the Universe had inflated, quarks formed the nuclei of the elements, and with electrons, atoms were formed.

**ELECTRON**
Negatively charged elementary particle.

**PHOTON**
Light elementary particle with no mass.

**GRAVITON**
A particle believed to transfer gravity.

**GLUON**
Responsible for interactions between quarks.

**QUARK**
Light elementary particle.

# THE COSMIC INFLATION THEORY

Big Bang theorists have been unable to understand with any level of certainty why the Universe has grown so quickly throughout its evolution. In 1979, physicist Alan Guth resolved this problem with his Inflation Theory. In an extremely short space of time (less than a thousandth of a second), the Universe grew 100 trillion trillion trillion trillion trillion times.

## HOW IT DIDN'T GROW

If there had been no inflation, the Universe would have comprised a series of clearly distinguishable regions. It would comprise 'remnants,' each of which would contain certain types of galaxy.

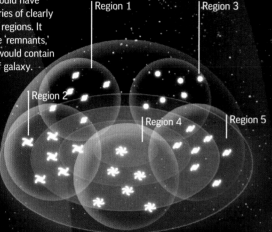

Region 1
Region 3
Region 2
Region 4
Region 5

## WMAP (WILKINSON MICROWAVE ANISOTROPY PROBE)

NASA's WMAP project makes it possible to see the Universe's background radiation. In the picture, hotter (red-yellow) and colder (green-blue) areas can be seen. WMAP makes it possible to establish the amount of dark matter.

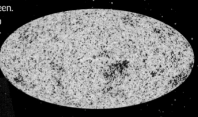

## SEPARATION OF FORCES

Before the inflation, there was just a single force that governed all interactions. The first to separate was gravity, then electromagnetic force and finally nuclear interaction. As the forces separated, matter was created.

GRAVITY
STRONG NUCLEAR
WEAK NUCLEAR
ELECTROMAGNETIC FORCE
SUPERFORCE
INFLATION

---

## $10^{-12}$ SECONDS

**$10^{15°}$ C**

The Universe experiences an immense cooling process. Gravity becomes separated, electromagnetic force appears and nuclear interaction begins.

## $10^{-4}$ SECONDS

(5) **$10^{12°}$ C**

Protons and neutrons are born, formed by three quarks each. The Universe is still dark: light is trapped in the mass of particles.

## 5 SECONDS

(6) **$5 \times 10^{9°}$ C**

Electrons and positrons annihilate one another, until the positrons disappear. The remaining electrons go on to form atoms.

## 3 MINUTES

(7) **$10^{9°}$ C**

They create the nuclei of the lightest elements: helium and hydrogen. Each nucleus comprises protons and neutrons.

**1 SEC** NEUTRINOS DECOUPLE AS A RESULT OF NEUTRON DISINTEGRATION. WITH A VERY SMALL MASS, NEUTRINOS GO ON TO FORM MOST OF THE UNIVERSE'S DARK MATTER.

### FROM PARTICLE TO MATTER

Quarks interacted with others thanks to the force transferred by gluons. Later, along with neutrons, they formed nuclei.

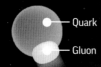

— Quark
— Gluon

(1) A gluon interacts with a quark.

(2) Quarks join gluons to form protons and neutrons.

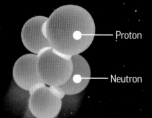

— Proton
— Neutron

(3) Protons and neutrons join to form nuclei.

# THE TRANSPARENT UNIVERSE

The creation of atoms and general cooling allowed the Universe, which was dense and opaque, to become transparent. Photons, light particles with no mass, were free to roam space; radiation lost its crown as governor of the Universe and matter was able to carve out its own destiny under the forces of gravity. Gaseous accumulations grew, and over hundreds of millions of years formed protogalaxies; thanks to gravity, they became the first galaxies, and in denser regions, the first stars began to burn brightly. The great, lingering mystery was why the galaxies took on their current shape. Dark matter – an intergalactic empty space – could hold the key; it was responsible for their expansion and can only be detected indirectly.

① **Gas cloud**
The first gases and dust generated by the Big Bang formed a cloud.

② **First filaments**
As a result of dark matter's gravity, gases joined together to form filaments.

## DARK MATTER

Dark matter, invisible to the most powerful telescopes, comprises 22 percent of the Universe's matter. Galaxies and stars move as a result of the gravitational effects of dark energy and matter.

## EVOLUTION OF MATTER

The Big Bang initially produced a gas cloud that was uniformly dispersed. Three million years later, the gas started to organize itself into the shape of filaments. Today, the Universe can be seen as networks of galactic filaments with enormous spaces between them.

| TIME | 380,000 YEARS AGO | 500 MILLION |
|---|---|---|

| TEMPERATURE | ⑧ 2,700° C (4,892° F) | ⑨ -243° C (-405° F) |
|---|---|---|

**⑧** Atoms are born. Electrons orbit around the nuclei, attracted by protons. The Universe becomes transparent. Photons travel throughout space.

**⑨** Galaxies acquire their definitive shape: 'islands' with billions of stars and masses of gas and dust. Stars explode, as supernovas, and scatter heavier elements such as carbon.

## FIRST ATOMS

Helium and hydrogen were the first elements to be joined at an atomic level. They are the main components of stars and planets, and the most common throughout the Universe.

NUCLEUS 1 | Proton

Electron

Neutron

NUCLEUS 2

① **Hydrogen**
An electron is attracted and orbits around a nucleus, which contains a proton and a neutron.

② **Helium**
As the nucleus has two protons, two electrons are attracted.

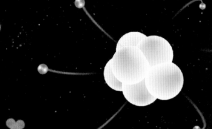

③ **Carbon**
Over time, more complex elements, like carbon, were formed (six protons and six electrons).

**Networks of filaments**
The Universe can be
seen as filaments with
billions of galaxies.

## THE UNIVERSE TODAY

Irregular
galaxy

Spiral
galaxy

Barred
spiral
galaxy

Elliptical
galaxy

Nebula

Star

Quasar

Galaxy
cluster

# 9.1 BILLION

**THE EARTH IS CREATED. LIKE ALL THE OTHER PLANETS, THE EARTH WAS FORMED FROM MATERIAL LEFT OVER FROM THE CREATION OF THE SUN.**

## 9 BILLION

## 13.7 BILLION

⑩ **-258° C (-432° F)**
Nine billion years after the Big Bang, the Solar System is
born. A mass of gas and dust collapses, giving rise to the
creation of the Sun. Then, with the remaining material, a
planetary system is brought together.

⑪ **-270° C (-454° F)**
Currently, the Universe continues expanding
with a countless number of galaxies separated by
dark matter. The predominant energy is also an
unknown: dark energy (74 percent).

## COSMIC CALENDAR

In an attempt to make the magnitudes of time related to the Universe more tangible, US
writer Carl Sagan introduced the concept of the 'Cosmic Calendar.' On 1 January of that
imaginary year, at 00.00 a.m., the Big Bang occurred. Homo sapiens would appear at 11.56
p.m. on 31 December, with Columbus discovering America (1492), at 11:59 a.m., that same
day. One second in the Cosmic Calendar is representative of 500 years.

**BIG BANG**
Occurring on the first second,
of the first day of the year.
**JANUARY**

**THE SOLAR SYSTEM**
Created on 24 August of
the Cosmic Calendar.

**COLUMBUS ARRIVES IN AMERICA**
This would occur on the last second of 31
December.

**DECEMBER**

# FORCES OF THE UNIVERSE

**The four main forces that inhabit space are those that cannot be explained using more basic forces.** Each one participates in different processes and each interaction involves different types of particles. Gravity, electromagnetic force, strong nuclear force and weak nuclear force are indispensable in the understanding of how the objects in the Universe behave.

## GENERAL THEORY OF RELATIVITY

The main contributions to understanding how the Universe works were formulated by Albert Einstein in 1915. Einstein thought of space as being connected to a dimension that nobody had previously considered: time. And gravity, which to Newton was the force that generated the attraction between two objects, was proposed by Einstein as the consequence of what he called the 'space-time curvature.' According to his theory of relativity, the Universe is curved by the presence of objects with different masses. And so, gravity is a spatial distortion which establishes that one object 'pulls' towards another, depending on whether the curve is greater or smaller.

Real position   What we see

LIGHT TRAJECTORY

Positive pole

SUN

Negative pole

$$E=MC^2$$

EARTH

**AS PART OF EINSTEIN'S EQUATION, ENERGY AND MASS ARE INTERCHANGEABLE. IF AN OBJECT INCREASES ITS MASS, THE ENERGY IT EMITS ALSO INCREASES.**

① GRAVITY

**The first to separate itself from the original super force. It is an attraction force that is currently perceived as set out by Einstein: as an effect of the space-time curvature. If you were to think of the Universe as a cube, the presence of any object with mass would generate a deformation in that cube. Gravity has the special feature of being able to act from a significant distance (like electromagnetism); however, it always exerts an attraction force.**

If the Universe contained no objects with mass, it might look like this.

The Universe is permanently undergoing deformations due to the masses of objects.

## UNIVERSAL GRAVITATION

Gravitation, as proposed by Newton, is the mutual attraction between two bodies of certain masses. Newton's Law, a paradigm accepted until Einstein's time, did not take into account either time or space as an essential part of the interaction between two objects. Attraction was caused by mass: objects with a greater mass attract objects with an inferior mass. And this was attributable solely to the intrinsic nature of objects. Nonetheless, the Law of Universal Gravitation was the pillar of Einstein's theory.

## NEWTON'S EQUATION

Two bodies with different masses are attracted. The body with the greater mass attracts the lighter body. The farther the distance between them, the lower the force.

$$F = \frac{GM_1 M_2}{D^2}$$

M₁ —— F —— M₂
D

## ③ STRONG NUCLEAR FORCE

It keeps the components of atomic nuclei together. Gluons are the particles responsible for transporting strong nuclear force and their immensity allows quarks to join together to form nuclear particles: protons and neutrons.

### QUARKS AND GLUONS

The strong interaction is transmitted when the gluon interacts with the quarks.

Nucleus

Quark
Force
Gluon

### UNION

Quarks join together and form nuclear protons and neutrons.

## ② ELECTROMAGNETIC FORCE

The force that affects electrically charged bodies. It participates in the chemical and physical transformations of atoms and molecules that form part of different elements. It is more intense than gravitational force and is active on two fronts, or poles: positive and negative.

### ATTRACTION

Two atoms are attracted and the electrons rotate around the new molecule.

Hydrogen

Helium

Force
Electron
Positive pole
Nucleus
Negative pole

### MOLECULAR MAGNETISM

Electromagnetic force is the predominant force in atoms and molecules. It is responsible for electrons orbiting around the nucleus, in light of their attraction to protons. The same happens between charged atoms that attract one another.

### LIGHT CURVES

Light also curves due to the space-time curvature. Seen from a telescope, the real position of an object is distorted. What the telescope sees is a false location, generated by the curve of light. It is not possible to witness the true position of the object.

## ④ WEAK NUCLEAR FORCE

The least intense force compared with all other forces. A weak interaction is active in the disintegration of a neutron, during which a proton and a neutrino are released; this later transforms into an electron. This force is active in natural radioactive phenomena, which occur in the atoms of certain particles.

### HYDROGEN

A hydrogen atom interacts with a light and weak particle (Wimp). A Down quark in the neutron becomes an Up quark.

### HYDROGEN ATOM

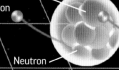

Proton
Electron
Neutron

### HELIUM ISOTOPE

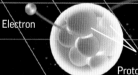

Electron
Proton

### HELIUM

The neutron becomes a proton. An electron is released and a helium isotope is created, with no nuclear neutrons.

# THEORIES ABOUT THE FUTURE

**To predict the future of the Universe, its total mass first needs to be discovered; to this day, this piece of data has eluded humankind.** According to the latest observations of astronomers, it is likely that the Universe's mass is significantly lower than the mass required to slow down its expansion. Thus, the current rate of growth is just a stepping stone on the path to complete destruction – and then total darkness.

## ① FLAT UNIVERSE

**With a mass equal in size to the critical point, the Universe would grow at a decreasing rate, but it would not reach a complete stop. The consequence of this would be the existence of an infinite number of galaxies and stars. If the Universe were flat, it would never end.**

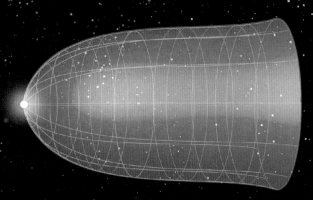

① The Universe evolves in constant expansion.

② It is constantly growing, but at an increasingly slower speed.

③ Gravity is not enough to create a total stop.

④ The Universe expands infinitely.

## HAWKING'S UNIVERSE

The Universe originally comprised four spatial dimensions, but none were temporary. As without time there is no change; one of these dimensions, according to Stephen Hawking, spontaneously transformed, and on a small scale, in the time dimension. And the Universe started to expand.

Object in three dimensions

Object that changes over time

**BIG BANG**

① After the initial explosion, the Universe grows.

② A continued, significant level of expansion is observed.

## ② CLOSED UNIVERSE

**If the amount of mass in the Universe was above the critical point, it would continue expanding until gravity held it back. It would then contract until a 'Big Crunch' occurred – a complete collapse which would culminate in a small, dense and infinitely hot mass, such as the one from which it was created.**

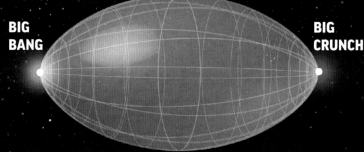

**BIG BANG**

**BIG CRUNCH**

## COMPOSITION

Although it is still unknown, the Universe's main energy source is dark energy.

**74 PERCENT**
Dark energy.

**22 PERCENT**
Dark matter.

**4 PERCENT**
Visible matter.

① The Universe violently expands due to its matter.

② It reaches a point at which growth starts to decrease.

③ The Universe collapses into a dense, hot unit.

## DISCOVERIES

The key discovery, which supported the existence of a Big Bang, was made by Edwin Hubble, who discovered that the galaxies are in constant expansion. Twenty years later, George Gamow proposed the existence of original background radiation. At the Bell Labs in New Jersey, Arno Penzias and Robert Wilson accidentally detected a constant signal throughout space at a temperature of -270° C: a fossil of the Universe's early radiation.

### 1920s EDWIN HUBBLE
He noted a deviation towards red on the spectrum and was able to establish that galaxies moved away from one another.

### 1940s GEORGE GAMOW
Russia's Gamow was the first to propose the Big Bang theory. He maintained that the early Universe was a 'melting pot' of particles.

### 1964 PENZIAS & WILSON
They discovered that, regardless of where they aimed, their antenna picked up a constant signal: background radiation.

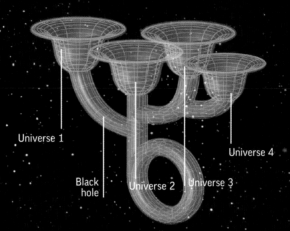

Universe 1

Black hole

Universe 2

Universe 3

Universe 4

## ③ SELF-REGENERATING UNIVERSES

A less commonly accepted theory is that universes generate themselves. In this instance, there would be several universes that are continuously recreating themselves. Self-regenerating universes could be transmitted by supermassive black holes.

③ A point arrives at which everything dies and life comes to an end.

## ④ OPEN UNIVERSE

The most accepted theory on the future of the cosmos is that the Universe's mass is lower than the critical point. The latest measurements seem to indicate that the current moment of expansion is just one phase prior to death. One day, the Universe will be extinguished for good.

### BLACK HOLES
It is believed that by travelling through a black hole, it may be possible to travel through space and get to know other universes. This would be possible due to antigravity effects.

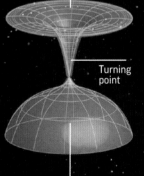

Black hole

Turning point

New universe

Universe 1

Universe 2

Universe 3

## ⑤ BABY UNIVERSES

According to this theory, universes continually give rise to other universes. However, in this case, a universe would be created after the death and disappearance of another universe, which would create a supermassive black hole; it would be from here that another universe is born. This process could be repeated indefinitely and it would be impossible to determine the number of universes in existence.

# ANATOMY OF THE GALAXIES

Galaxies are groups of stars, gas and dust that are constantly in rotation. The first were formed one hundred million years ago after the Big Bang; today, there are billions throughout space. They take on very different shapes, with the greatest number of stars accumulating at their core. Galaxies tend to group together in space due to the effects of gravity; in doing so, they form clusters of hundreds or thousands of galaxies, with varying and different forms.

## SOMBRERO GALAXY

This galaxy is located 28 million light years from Earth and its name is attributable to the special shape of its spiral arms, which encompass a shining, white core.

⌄

## GALACTIC COLLISIONS

In 1926, scientist Edwin Hubble proposed the existence of faraway galaxies; just three years later, he confirmed that they were moving away from the Milky Way, demonstrating that the Universe was constantly expanding. However, galaxies tend to find one another and 'galactic collisions' cause galaxies to merge and result in a clash over gaseous matter. The future of the Universe will be made up of fewer, much larger and much more dense galaxies.

① **1.2 billion years ago,** the Antennae Galaxies (NGC 4038 and NGC 4039) were two separate spiral galaxies.

② **300 million years later,** the galaxies start to interact; they charge towards one another at high speed.

③ **300 million years** pass until a mutual cross is generated, as part of which the shapes of the galaxies change.

# HUBBLE'S CLASSIFICATION OF THE GALAXIES

## ELLIPTICAL GALAXIES
Galaxies with old stars in the form of a sphere. They contain a small amount of dust and gas. Their masses vary in size.

## SPIRAL GALAXIES
A core of old stars is encompassed by a flat disc of stars with two or more spiral arms.

## IRREGULAR GALAXIES
Galaxies with no defined form that cannot be classified. They are abundant in gas and dust clouds.

| EO | E3 | E5 | E7 | Sa | Sb | Sc |

## SUBCLASSIFICATIONS
Galaxies are subdivided into different categories, depending on whether they are more or less circular (in the case of elliptical galaxies, that range from EO to E7) and depending on whether their arms and axis are larger or smaller (in the case of spiral galaxies, that run from Sa to Sc).
An EO galaxy is an almost circular galaxy, while an E7 has a flatter, oval shape. An SA galaxy has a large central axis and distinctly curved arms, whereas an SC has a smaller axis and more spread-out arms.

# CLUSTERS
Galaxies are objects that tend to form groups or clusters. Acting in response to gravitational force, they can form clusters of galaxies of anywhere from two to thousands of galaxies. These clusters have various shapes and are thought to expand when they join together. Abell 2151 (The Hercules cluster), shown here, is located approximately 500 million light-years from Earth. Each dot represents a galaxy that includes billions of stars.

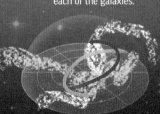

**300 million years later,** the stars of the spiral arms fly off each of the galaxies.

**In its present state,** two jets of expelled stars extend away from the original galaxies.

## COLLISION
Located 300 light years away from Earth, these two colliding galaxies form part of the so-called 'Mice' galaxies; their name is attributable to their large tails of stars and gas that emanate from each galaxy. Over time, they will fuse into a single, larger galaxy.

# OUR GALAXY: THE MILKY WAY

**For a long time our galaxy, the Milky Way, named as such due to its milky, bandlike appearance, was a true mystery.** It was Galileo Galilei who, in 1610, directed his telescope and saw that the weak cloudy-white band was comprised of thousands and thousands of stars, practically stuck to one another. Gradually, astronomers started to realize that all the stars, and our Sun, formed part of one large entity: a galaxy, our huge stellar home.

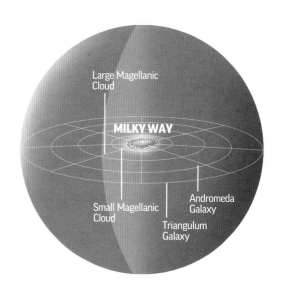

Large Magellanic Cloud

**MILKY WAY**

Small Magellanic Cloud

Triangulum Galaxy

Andromeda Galaxy

## STRUCTURE OF THE GALAXY

Our galaxy has two spiral arms that rotate around the core. It is on these arms that the youngest objects in our galaxy can be found, and where interstellar gas and dust are most abundant. On the Sagittarius Arm, one of the brightest stars in the Universe can be found – Eta Carinae. Our Solar System is located on the inner border of the Orion Arm, between the Sagittarius and Perseus Arms.

### ROTATION
The speed at which the Milky Way rotates varies depending on the distance from the core of the galaxy. Between the core and edge, where most stars can be found, the speed is much greater as objects feel the attraction of billions of stars.

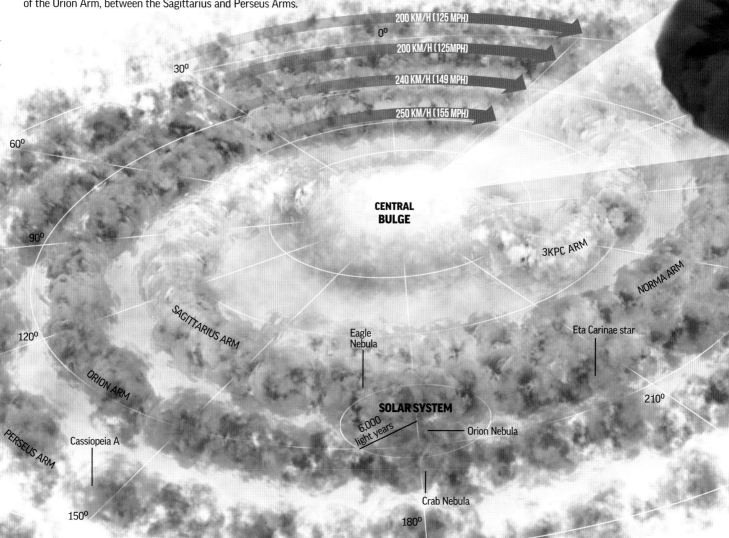

200 KM/H (125 MPH)

0°

200 KM/H (125MPH)

30°

240 KM/H (149 MPH)

250 KM/H (155 MPH)

60°

90°

120°

150°

180°

210°

CENTRAL BULGE

3KPC ARM

NORMA ARM

SAGITTARIUS ARM

ORION ARM

PERSEUS ARM

Cassiopeia A

Eagle Nebula

Eta Carinae star

SOLAR SYSTEM

6,000 light years

Orion Nebula

Crab Nebula

# CENTRAL REGION

The central axis of the galaxy contains old stars, dating back around 14 billion years, and exhibits an intense level of activity at its interior. Here, two hot gas clouds can be found: Sagittarius A and B. In the central region, although outside the core, a giant cloud contains 70 different types of molecule. These gas clouds are attributable to violent activities at the centre of the galaxy. The heart of the Milky Way can be found in the depths of Sagittarius A and B.

**MAGNETISM**
The heart of the galaxy is encompassed by a region with strong magnetic fields, perhaps resulting from a rotating black hole.

**SAGITTARIUS B2**
The largest dark cloud in the central area.

**OUTER RING**
A ring of smoke and dark dust and molecule clouds expand as a result of a huge explosion. It is suspected that this was attributable to a small object towards the center.

CARINA ARM

240°

OUTER ARM

# 200 BILLION
**STARS INHABIT THE MILKY WAY. THE NUMBER OF SUNS IT CONTAINS IS SO HIGH THAT THEY CANNOT BE SEPARATED.**

**HOT GASES**
Emitted from the surface of the central part. They may be the result of violent explosions on the accretion disc.

**BLACK HOLE**
One is believed to occupy the center of the galaxy. It attracts gas due to its gravitational pull and keeps it in orbit.

**SHINING STARS**
They are created from gas that is not swallowed by the black hole. Most are young.

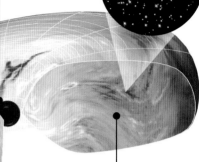

**GASEOUS VORTEXES**
From the center outwards, the gas is stuck and concentrated by a gravitational force that may be attributable to a black hole.

# THE EXACT CENTER

The core of the Milky Way galaxy is marked by very intense radio-wave activity that might be produced by an accretion disk made up of incandescent gas surrounding a massive black hole. The region of Sagittarius A is a gas ring that rotates at very high speed, swirling within several light-years of the center of the galaxy. The speed of its rotation is an indication of the powerful gravitational force exerted from the center of the Milky Way.

# A DIVERSE GALAXY

The brightest portion of the Milky Way that appears in photographs taken with optical lenses is in the constellation Sagittarius, which appears to lie in the direction of the center of the Milky Way. In some cases, stars are obscured by dense dust clouds that make some regions of the Milky Way seem truly dark. The objects that can be found in the Milky Way are not all of one type. Some, such as those known as the halo population, are old and are distributed within a sphere around the galaxy. Other objects form a more flattened structure called the disk population. In the spiral arm population, we find the youngest objects in the Milky Way.

**SAGITTARIUS**
Close to the center of the Milky Way, Sagittarius shines intensely.

**DARK REGIONS**
Dark regions are produced by dense clouds that obscure the light of stars.

**STARS**
So many stars compose the Milky Way that it is impossible for us to distinguish them all.

**THE MILKY WAY IN VISIBLE LIGHT**

**SECTORS**
Many different sectors make up the Milky Way.

# ACTIVE GALAXIES

**There is a small proportion of galaxies that are different from the others, distinguished by their high level of energetic activity.** This could be attributable to the presence of black holes at their core, formed as a result of the death of supermassive stars. It is very likely that these cores of the first galaxies are 'quasars', which can be seen today at a remote distance.

## GAS

Two jets are expelled from the core and emit radio waves. If they cross paths with intergalactic gas clouds, they form huge clouds capable of emitting X-rays or radio waves.

## ENERGETIC ACTIVITY

It is believed that active galaxies are a direct descendent of the early stages of the Universe. After the Big Bang, these galaxies would have been left with a significant amount of energetic radiation. Quasars, small, dense and luminous, comprise the cores of this type of galaxy. In certain cases, X-rays may be emitted, but in others, radio waves.

## CENTRAL RING

The core is covered by a ring of dust and gas, dark on the inside and shiny on the outside. It is a potent energy source.

## THE FORCE OF GRAVITY

Vast amounts of hot gas clouds start to join together. The clouds attract one another and collide. As part of these collisions, stars are created. A large amount of gas accumulates at the center of the galaxy. The gravitational force increases until it reaches such a level of intensity that a massive black hole grows at the core.

## THE CORE'S QUASAR

It expels two jets of particles that reach a speed close to the speed of light. The gas and stars generated by the jets sent into space are swallowed in the form of a spiral by the black hole, creating an accretion disc and a quasar.

## BLACK HOLE

It swallows all gas that starts to surround it. It forms a gaseous, hot spiral that also emits jets at high speed. Its magnetic field dumps charged particles around the black hole. The outside of the disc feeds on interstellar gas.

## CLASSIFICATION

The classification of an active galaxy depends upon its distance from Earth and the perspective from which it is seen. Quasars, radio galaxies, and blazars are members of the same family of objects and differ only in the way they are perceived.

### QUASARS
The most powerful objects in the universe, quasars are so distant from Earth that they appear to us as diffuse stars. They are the bright cores of remote galaxies.

### RADIO GALAXIES
Radio galaxies are the largest objects in the universe. Jets of gases come out from their centers that extend thousands of light-years. The cores of radio galaxies cannot be seen.

### BLAZARS
Blazars may be active galaxies with jets of gas that are aimed directly toward Earth. The brightness of a blazar varies from day to day.

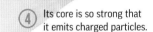

④ Its core is so strong that it emits charged particles.

③ The disc's strong gravitational pull attracts and destroys stars.

### PARTICLES
Expelled from the black hole, they contain intense magnetic fields and charged particles. The jets travel at speeds close to the speed of light.

### ACCRETION DISC
Formed by interstellar gas and the remnants of stars, they are capable of emitting X-rays given the extreme levels of heat at their centre.

## GALAXY FORMATION

A theory of galaxy formation associated with active galaxies holds that many galaxies, possibly including the Milky Way, were formed from the gradual calming of a quasar at their core. As the surrounding gases consolidated in the formation of stars, the quasars, having no more gases to absorb, lost their energetic fury and became inactive. According to this theory, there is a natural progression from quasars to active galaxies to the common galaxies of today.

### GAS CLOUDS
Formed as a result of the gravitational collapse of immense masses of gas during the first stages of the Universe. Later, in their interior, stars were formed.

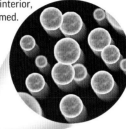

### INCREASING GRAVITY

① Dark gas and dust clouds are located on its outer edge. They are gradually swallowed by the hole.

② The gas moves inwards, gradually heating up.

## 100 MILLION
**DEGREES CELSIUS (180 MILLION º F) – THE TEMPERATURE THAT CAN BE REACHED AT THE CORE OF A BLACK HOLE.**

④

## STABLE GALAXIES

**It is commonly accepted that most galaxies are formed from the progressive inactivity of nuclear quasars. As gases gather together to form stars, the quasars are left with no further gas left to swallow and so they are rendered inactive.**

# WITH THEIR OWN LIGHT

**For a long time, stars have been a mystery to humankind.** Today, it is known that they are enormous spheres of incandescent gases, mostly hydrogen, with a smaller amount of helium. Based on the light they emit, experts can ascertain their brightness, color and temperature. Given their huge distance from Earth, they can only be seen as dots of light, even with the most powerful telescopes.

## HERTZSPRUNG-RUSSELL DIAGRAM

The H-R diagram groups stars according to their visual brightness, the spectral type that corresponds to the wavelengths of light they emit and their temperature. Stars with a greater mass are brighter – such as blues, red giants and red supergiants. Stars live 90 percent of their lives in the so-called 'main sequence.'

**O-TYPE** (40,000 to 29,000° C)

**B-TYPE** (29,000 to 9,700° C)

**A-TYPE** (9,700 to 7,200° C)

**F-TYPE** (7,200 to 5,800° C)

**G-TYPE** (5,800 to 4,700° C)

**K-TYPE** (4,700 to 3,300° C)

**M-TYPE** (3,300 to 2,100° C)

## LIGHT YEARS AND PARSECS

In order to measure the immense distance between stars, the terms light year (ly) and parsec (pc) are used. A light year is the distance light travels in one year: almost 10 trillion km (6.2 trillion miles). A pc is the distance between a star and Earth, if its parallax angle is one arc-second. One pc equals 3.26 ly, or 31 trillion km (19.3 trillion miles).

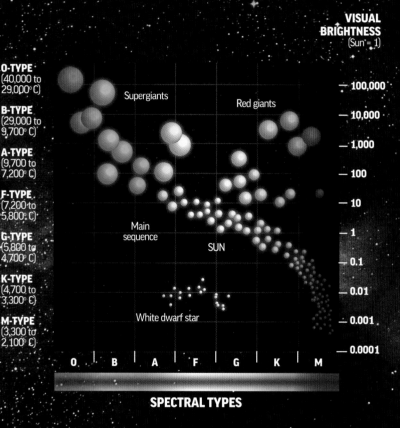

VISUAL BRIGHTNESS (Sun = 1)

Supergiants

Red giants

Main sequence

SUN

White dwarf star

— 100,000
— 10,000
— 1,000
— 100
— 10
— 1
— 0.1
— 0.01
— 0.001
— 0.0001

O | B | A | F | G | K | M

**SPECTRAL TYPES**

**MAIN STARS, LESS THAN 100 LIGHT YEARS FROM THE SUN**

**SUN** (G2)

**ALPHA CENTAURI** (G2, K1, M5)

**SIRIUS** (A0 and dwarf star)

**PROCYON** (F5 and dwarf star)

**ALTAIR** (A7)

**VEGA** (A0)

**POLLUX** (K0 giant)

**ARCTURUS** (K2 giant)

**CAPELLA** (G6 and G2 giants)

LIGHT YEARS

0 1 2 3 4 5 6 7 8 9 10 11 12 13 14 15 16 17 18 19 20 21 22 23 24 25 26 27 28 29 30 31 32 33 34 35 36 37 38 39 40 41 42 43 44 45 46 47 48 49

0 PARSECS 1    2    3    4    5    6    7    8    9    10    11    12    13    14    15

# MEASURING DISTANCE

When the Earth orbits the Sun, the closest stars appear to move over a background of more distant stars. The angle that results from the movement of a star within the Earth's six-month rotation period is known as the 'parallax angle'. The closer a star is to Earth, the larger the parallax.

The parallax angle of star A is small. Therefore, it is a long way from Earth.

The parallax angle of star B is greater than that of A. B is therefore closer to Earth.

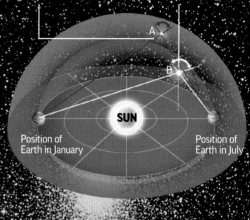

A

B

SUN

Position of Earth in January

Position of Earth in July

## COLORS

The hottest stars are blue-white (O-, B- and A-spectral type). The coldest, G-, K- and M-type stars, are orange, yellow and red.

## GLOBULAR CLUSTER

Around 10 million stars join together to form a huge band: Omega Centauri.

## OPEN CLUSTER

The Pleiades are a formation of around 400 stars that will scatter throughout space in the future.

**CASTOR**
(A2, A1 and M1)

**ALDEBARAN**
(K5 giant)

**ALIOTH**
(A0 giant)

**REGULUS**
(B7 and K1)

**MENKALINAN**
(A2 and A2)

**GACRUX**
(M4 giant)

**ALGOL**
(B8 and K0)

52 53 54 55 56 57 58 59 60 61 62 63 64 65 66 67 68 69 70 71 72 73 74 75 76 77 78 79 80 81 82 83 84 85 86 87 88 89 90 91 92 93 94 95 96 97 98 99 100

16    17    18    19    20    21    22    23    24    25    26    27    28    29    30

# STELLAR EVOLUTION

**Stars are born in nebulae, enormous gas clouds, mainly comprised of hydrogen and dust that float in space.** They can live for millions, or even billions, of years. Often, their size can provide clues about their age: smaller stars tend to be younger, while larger stars are closer to perishing, soon to cool down or explode as supernovas.

## MASSIVE STAR
**More than eight solar masses.**

### ② STAR
The star is born. Hydrogen fuses to form helium during the main sequence.

### ① PROTOSTAR

Comprised of a dense, gaseous core surrounded by a dust cloud.

### NEBULA
A cloud of gas and dust collapses due to the effects of gravity; it heats up and divides into smaller clouds. Each of these clouds forms a 'protostar.'

### ② STAR

It shines and slowly consumes its hydrogen reserves. It fuses helium while it grows in size.

## SMALL STAR
**Less than eight solar masses.**

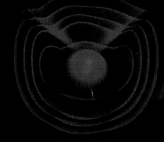

## THE LIFE CYCLE OF A STAR

The evolution of a star depends on its mass. Smaller stars, like the Sun, have much longer, more modest lives. When they run out of hydrogen, they turn into a red giant and, eventually, end their lives as white dwarves until they completely burn out. Stars with a greater mass eventually explode: all that is left of them is a hyper-dense remnant – neutron stars. Significantly, more massive stars eventually form black holes.

### ① PROTOSTAR

Formed by the release of gas and dust. Its core turns due to the effects of gravity.

### ③ RED SUPERGIANT

The star expands and heats up, forming a heavy iron core.

## 95% OF STARS

END THEIR LIVES AS WHITE DWARF STARS. OTHER, LARGER STARS EXPLODE AS SUPERNOVAS ILLUMINATING ENTIRE GALAXIES FOR WEEKS.

### ④ SUPERNOVA

When the star is no longer able to fuse more elements, the core collapses, leading to a large release of energy.

### ⑤ BLACK HOLE

If the initial mass is greater than 20 suns, the core is even denser and forms a black hole, with a highly intense gravitational pull.

### ⑤ NEUTRON STAR

If the initial mass varies between eight and twenty suns, the resulting star is a neutron star.

### ⑥ BLACK DWARF STAR

If it completely burns out, the white dwarf star turns into a black dwarf star. They cannot be seen in space.

### ③ RED GIANT

The star continues to grow. The core heats up. As it runs out of helium, it fuses carbon and oxygen.

### ⑤ WHITE DWARF STAR

The star is encompassed by the gases and loses brightness.

### ④ PLANETARY NEBULA

Having run out of fuel, the core condenses and the outer layers are detached. The gases released form gas clouds.

# RED GIANTS

**When a star exhausts its hydrogen reserves, it starts to die.** When this happens, the core turns into a ball of helium and reactions begin to cease. The helium stays bright and luminous, until it consumes itself and the core contracts. The outer layers of the star dilate until it becomes a red giant, before finally cooling down.

## SPECTACULAR DIMENSIONS

When the star exhausts its hydrogen, it grows to 200 times the diameter of the Sun. It then starts to burn helium and its size decreases until it measures between 10 and 100 times the Sun. Its growth stabilizes at this point until, after billions of years, it dies. In the case of supergiants, they collapse before exploding.

**RED SUPERGIANT**
Placed at the center of the Solar System, it would swallow up Mars and Jupiter.

**RED GIANT**
Placed at the center of the solar system, it could reach only the nearer planets, such as Mercury, Venus, and the Earth.

- Sun
- Mercury's orbit
- Venus' orbit
- Earth's orbit
- Mars' orbit
- Jupiter's orbit
- Saturn's orbit

### CORE REGION
The core of a giant red is ten times smaller than the original core, as it shrinks due to a lack of hydrogen.

(1) **Hydrogen** It keeps burning outside the core when the core has run out of hydrogen.

(2) **Helium** This is produced after the hydrogen is burnt.

(3) **Carbon and oxygen** Produced during helium combustion, they fuse at the core of the red giant.

(4) **Temperature** While the helium burns, the core reaches 100 million degrees Celsius.

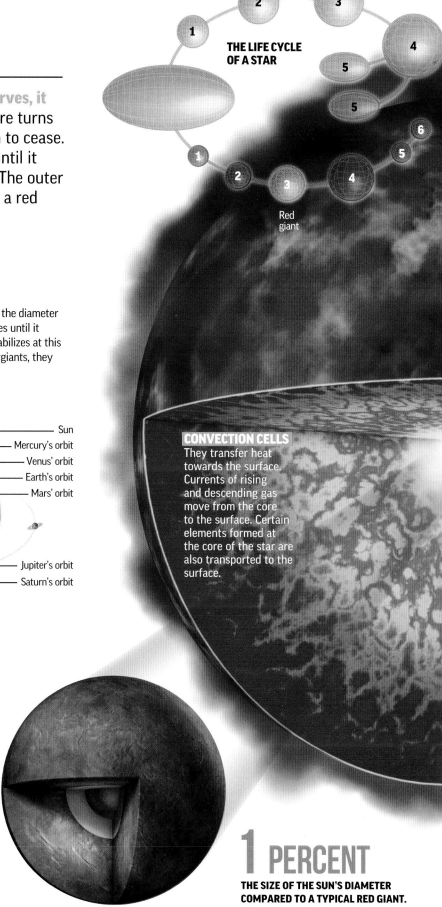

**THE LIFE CYCLE OF A STAR**

1  2  3  4  5  5  6

1  2  3  4  5

Red giant

**CONVECTION CELLS**
They transfer heat towards the surface. Currents of rising and descending gas move from the core to the surface. Certain elements formed at the core of the star are also transported to the surface.

# 1 PERCENT

**THE SIZE OF THE SUN'S DIAMETER COMPARED TO A TYPICAL RED GIANT.**

# 6 BILLION YEARS

**THIS IS HOW LONG IT WILL TAKE THE SUN TO SWALLOW UP THE EARTH.**

**HEAT SPOTS**
These appear when large currents of incandescent gas reach the surface. They can be seen on the surface of neighboring red giants.

**GRAINS OF DUST**
They condense in the outer atmosphere and then scatter by means of stellar winds. The dust is dispersed throughout interstellar space, where new generations of stars are formed.

## WHITE DWARF STAR

After experiencing a period as a red giant, stars like the Sun lose their outer layers, giving rise to a planetary nebula. At its heart is a white dwarf star, an extremely hot (200,000° C/360,000° F) and dense object. When it completely extinguishes, it becomes a black dwarf star.

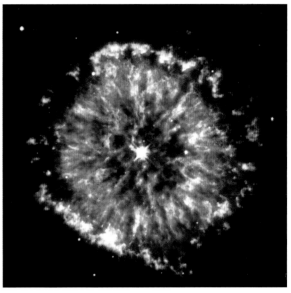

## NEBULA NGC 6751

**After the nuclear reaction in the star's core ceases, the star ejects its outer layers, which then form a planetary nebula.**

## THE FUTURE OF THE SUN

Like any typical star, the Sun burns hydrogen. It will take another five billion years to exhaust its reserves; only then will it become a red giant. Its brightness will multiply, and it will expand until it swallows Mercury and even Earth. Once it stabilizes, it will remain as a giant for two billion years, until it turns into a white dwarf star.

④ **Red giant**
The Sun's radius reaches the Earth's orbit.

300 million km (186 million miles)

**SUN**

Earth

① Venus Sun
Mars
Earth Mercury

② Venus Sun
Mars
Earth Mercury

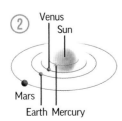

③ Venus Sun
Mars
Earth

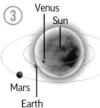

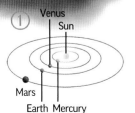

# GAS SHELLS

**When small stars die, all that is left of them are enormous shells of expanding gas;** these are known as 'planetary nebulae.' In general, they are symmetrical and spherical objects. When viewed through a telescope, a white dwarf star can be seen at the center of several nebulae – a remnant of the original star.

**THE LIFE CYCLE OF A STAR**

Planetary nebula

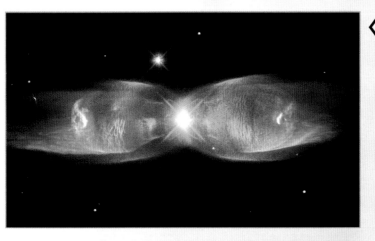

## M2-9

The Butterfly Nebula contains two stars that orbit around one another within a disc of gas that measures 10 times the size of the orbit of Pluto. It is 2,100 light years from Earth.

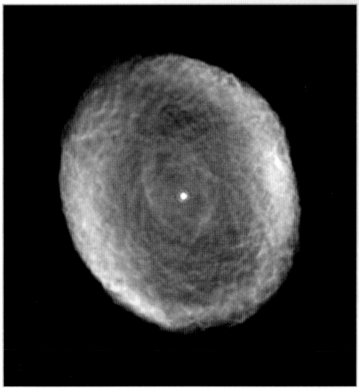

## IC 418

The Spirograph Nebula has a hot and bright core that stimulated neighboring atoms, making them glow. Located 2,000 light years from Earth, it measures 0.3 light years in diameter.

## HYDROGEN

Constantly expanding gaseous masses contain hydrogen for the most part, in addition to helium, and to a lesser extent, oxygen, nitrogen and other elements.

## TWICE

THE TEMPERATURE OF THE SUN – THE SURFACE TEMPERATURE OF A WHITE DWARF STAR. THAT IS WHY THEY APPEAR WHITE, DESPITE THEIR BRIGHTNESS BEING A THOUSAND TIMES SMALLER.

# 3 TONS

**EQUALS THE WEIGHT OF ONE SPOONFUL OF A WHITE DWARF STAR. ALTHOUGH THE SIZE IS SIMILAR TO THE SPOONFUL OF CREAM, ITS MASS IS VASTLY DIFFERENT. WHILE THE SPOONFUL OF CREAM WEIGHS VERY LITTLE, THE WEIGHT OF ONE SPOONFUL OF WHITE DWARF STAR IS 3 TONS. THE MASS OF A WHITE STAR IS IMMENSE, DESPITE ITS DIAMETER (15,000 KM/ 9,320 MI) BEING COMPARABLE TO THE EARTH'S DIAMETER.**

### CONCENTRIC CIRCLES

Circles of gas form an onion-layer structure around the white dwarf. The mass of each white dwarf is greater than all the masses of the Solar System's planets combined.

## NGC 7293

**The Helix is a planetary nebula created at the end of the life of a star similar to our Sun. It is 650 light years from Earth.**

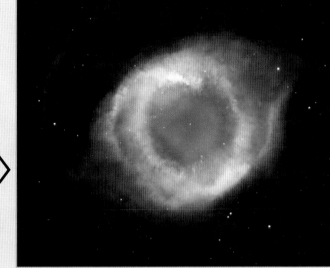

## MYCN 18

**Two rings of gas form the silhouette of the Hourglass Nebula. The red color corresponds to nitrogen, and the green color to hydrogen. This nebula is 8,000 light years from Earth.**

### WHITE DWARF STAR

The remainder of the red giant can be found at the heart of the nebula. The star cools down and, at some stage, completely extinguishes; as a result, it becomes a black dwarf star and can no longer be seen.

### GREATER DIAMETER
Less massive white dwarf

### SMALLER DIAMETER
More massive white dwarf

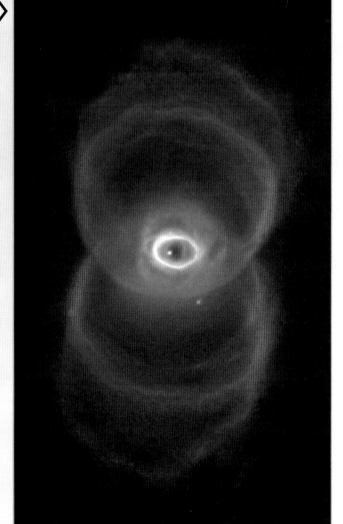

## DENSITY OF A WHITE DWARF

The density of a white dwarf is one million times greater than the density of water. In other words, one cubic meter (35.3 cu ft) of a white dwarf star would weigh one million tons. The mass of a star varies, and is indirectly proportional to its diameter. A white dwarf star, with a diameter 100 times smaller than the diameter of the Sun, has a mass 70 times greater.

# SUPERNOVA

**The explosion of stars towards the end of their lives is extraordinary;** there is a sudden increase in their brightness and an enormous release of energy. This is a 'supernova' and it releases, in just 10 seconds, 10 times more power than the Sun releases in its entire life. After the star's detonation, a gaseous remnant remains and this expands and shines for millions of years throughout the galaxy. It is estimated that two supernovas explode each century in the Milky Way.

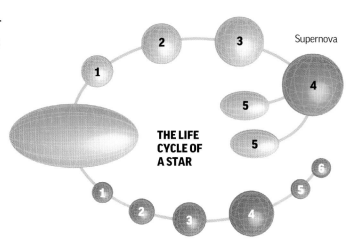

Supernova

**THE LIFE CYCLE OF A STAR**

## A STAR'S DECLINE

The explosion that sees the end of a supergiant's life is attributable to its extremely heavy iron core no longer being able to support its own gravitational pull. As internal fusions are no longer possible, the star collapses in on itself, expelling the remaining gases outwards; these then expand and shine for millennia. The expelled elements provide the interstellar medium with new material, which is capable of giving rise to new generations of stars.

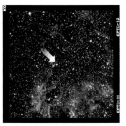

**BEFORE AND AFTER**

The image at left shows a sector of the Large Magellanic Cloud, an irregular galaxy located 170,000 light-years from the Earth, depicted before the explosion of supernova 1987A. The image at right shows the supernova.

## CORE

**Divided into different layers, each one corresponds to the different elements generated as a result of nuclear fusion. The final element created prior to the collapse is nuclear iron.**

## SUPERGIANT

**Once the star swells, it is capable of measuring more than 1,000 times the diameter of the Sun. The star is capable of producing elements heavier than carbon and oxygen.**

**FUSION**

The nuclear reactions happen more quickly than those of a red giant.

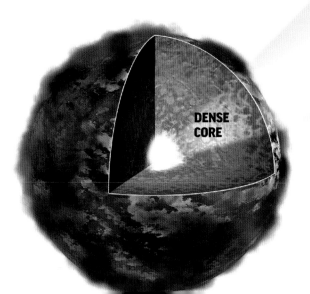

**DENSE CORE**

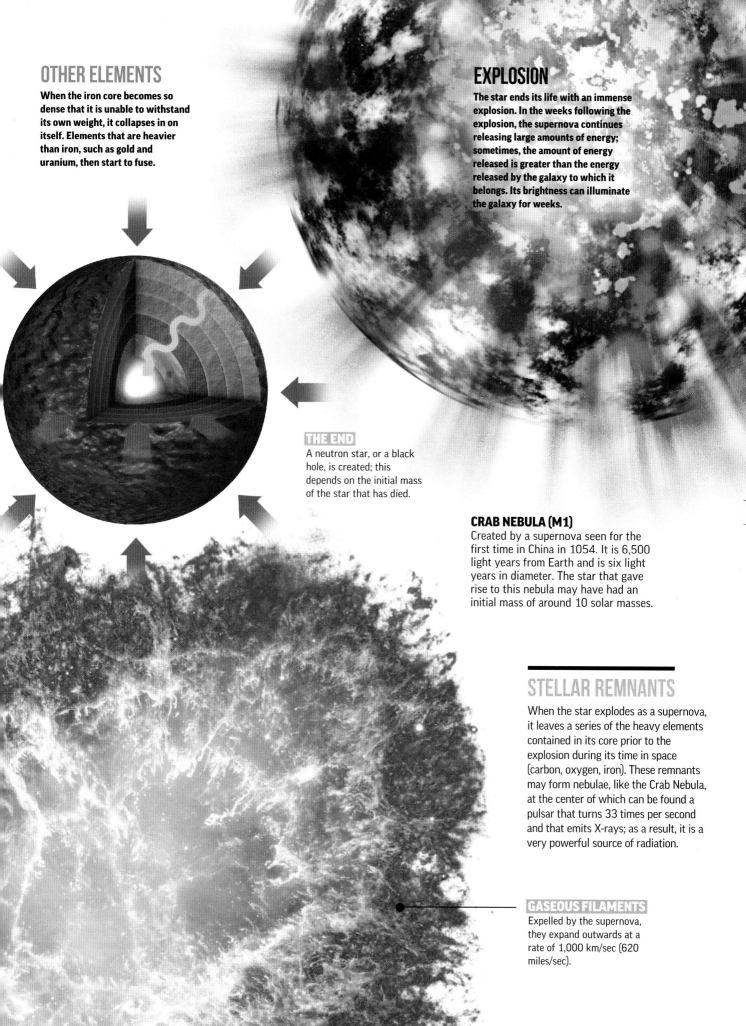

## OTHER ELEMENTS

When the iron core becomes so dense that it is unable to withstand its own weight, it collapses in on itself. Elements that are heavier than iron, such as gold and uranium, then start to fuse.

## EXPLOSION

The star ends its life with an immense explosion. In the weeks following the explosion, the supernova continues releasing large amounts of energy; sometimes, the amount of energy released is greater than the energy released by the galaxy to which it belongs. Its brightness can illuminate the galaxy for weeks.

### THE END

A neutron star, or a black hole, is created; this depends on the initial mass of the star that has died.

### CRAB NEBULA (M1)

Created by a supernova seen for the first time in China in 1054. It is 6,500 light years from Earth and is six light years in diameter. The star that gave rise to this nebula may have had an initial mass of around 10 solar masses.

## STELLAR REMNANTS

When the star explodes as a supernova, it leaves a series of the heavy elements contained in its core prior to the explosion during its time in space (carbon, oxygen, iron). These remnants may form nebulae, like the Crab Nebula, at the center of which can be found a pulsar that turns 33 times per second and that emits X-rays; as a result, it is a very powerful source of radiation.

### GASEOUS FILAMENTS

Expelled by the supernova, they expand outwards at a rate of 1,000 km/sec (620 miles/sec).

# THE FINAL DARKNESS

**The final stage in the evolution of a star's core is the formation of a very compact object.** The nature of which depends on the mass that collapses. Larger stars end up as black holes; these elements are so dense that they do not even allow light to pass through.

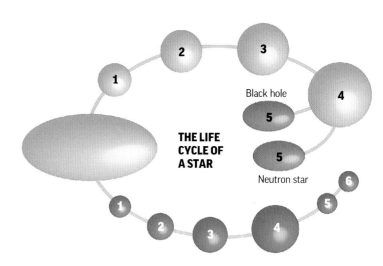

**THE LIFE CYCLE OF A STAR**

Black hole

Neutron star

## DISCOVERY OF BLACK HOLES

The only way of detecting the presence of a black hole in space is by its effect on neighboring stars. Since the gravitational force exerted by a black hole is so powerful, the gases of nearby stars are absorbed at great speed, spiraling toward the black hole and forming a structure called an accretion disk. The friction of the gases heats them until they shine brightly. The hottest parts of the accretion disk may reach 100,000,000° C and are a source of X-rays. The black hole, by exerting such powerful gravitational force, attracts everything that passes close to it, letting nothing escape. Since even light is not exempt from this phenomenon, black holes are opaque and invisible to even the most advanced telescopes.

### ACCRETION DISK
An accumulation of gases that a black hole absorbs from neighboring stars. The gas spins at an extremely high speed and, in areas very close to the black hole, results in the emission of X-rays.

**LIGHT RAYS**

**X-RAYS**
Gas enters the black hole and is heated up. This results in the emission of X-rays.

**FULL ESCAPE**
When light passes very far from the center, it follows its natural course.

**CLOSE TO THE LIMIT**
Light has still not crossed the event horizon, the limit between what is and what is not absorbed, and it maintains its brightness.

**DARKNESS**
Like any object, if it passes very close to the core, light is trapped.

### CROSS-SECTION
Accretion disc

X-rays

Igneous gases

Black hole

### SHINING GASES
As the accretion disc feeds on the gases that turn at extremely high speed, the part closest to the core shines intensely. Towards the edges, it has a colder and darker appearance.

# NEUTRON STAR

When the initial star has a mass of between 10 and 20 solar masses, its final mass will be greater than that of the Sun. Despite having lost significant amounts of matter during the nuclear reaction process, the star ends up with a very dense core, resulting in the creation of a 'neutron star.'

## LOSS OF MASS

Toward the end of its life, a neutron star loses more than 90 percent of its initial mass.

① **Red giant**
Its diameter is 100 times greater than the diameter of the Sun.

② **Supergiant**
It grows and quickly fuses elements. It produces carbon and oxygen until it finally forms iron.

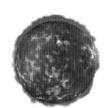

③ **Explosion**
The iron core collapses. Protons and electrons annihilate one another to form nuclear neutrons.

④ **Dense core**
Its exact composition is unknown. It contains interacting particles, most of which are neutrons.

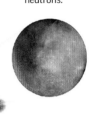

# 1 BILLION TONS

**THE WEIGHT OF ONE SPOONFUL OF A NEUTRON STAR. ALTHOUGH THE SIZE IS SIMILAR TO THE SPOONFUL OF PEANUT BUTTER, ITS MASS IS VASTLY DIFFERENT. WHILE THE SPOONFUL OF PEANUT BUTTER WEIGHS VERY LITTLE, THE WEIGHT OF ONE SPOONFUL OF NEUTRON STAR IS 1 BILLION TONS. IT HAS A COMPACT AND DENSE CORE, WITH AN INTENSE GRAVITATIONAL PULL.**

## CURVED SPACE

The theory of relativity suggests that gravity is not a force but a distortion of space. This distortion creates a gravitational well, the depth of which depends on the mass of the object. Objects are attracted to other objects through the curvature of space.

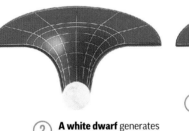

① **The Sun** forms a shallow gravitational well.

② **A white dwarf** generates a deeper gravitational well, drawing in objects at a higher speed.

③ **A neutron star** attracts objects at speeds approaching half the speed of light. The gravitational well is even more pronounced.

④ **Black hole**
The objects that approach the black hole too closely are swallowed by it. The black hole's gravitational well is infinite and traps matter and light forever. The event horizon describes the limit of what is, and is not, absorbed. Some scientists believe in the existence of so-called wormholes—antigravity tunnels, through which travel across the universe is hypothesized to be possible.

**WORMHOLE**

Entrance

Exit

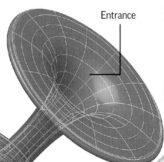

# PULSARS

The first pulsar, a neutron star that emits radio waves, was discovered in 1967. Pulsars turn 30 times per second and have a very intense magnetic field. Pulsars emit waves from both its poles while they spin. If it absorbs gas from a neighboring star, it generates a heat spot on its surface which emits X-rays.

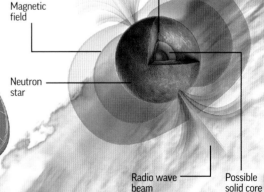

Rotation axis

Magnetic field

Neutron star

Radio wave beam

Possible solid core

## DEVOURING THE GAS OF A SUPERGIANT

Located in a binary system, pulsars are capable of following the same process as a black hole. Faced with a neighboring star with a smaller mass, due to its gravitational pull, it absorbs the star's gas, heating its surface. Thus, the pulsar is capable of emitting X-rays.

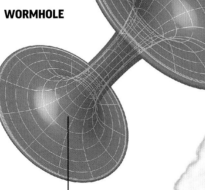

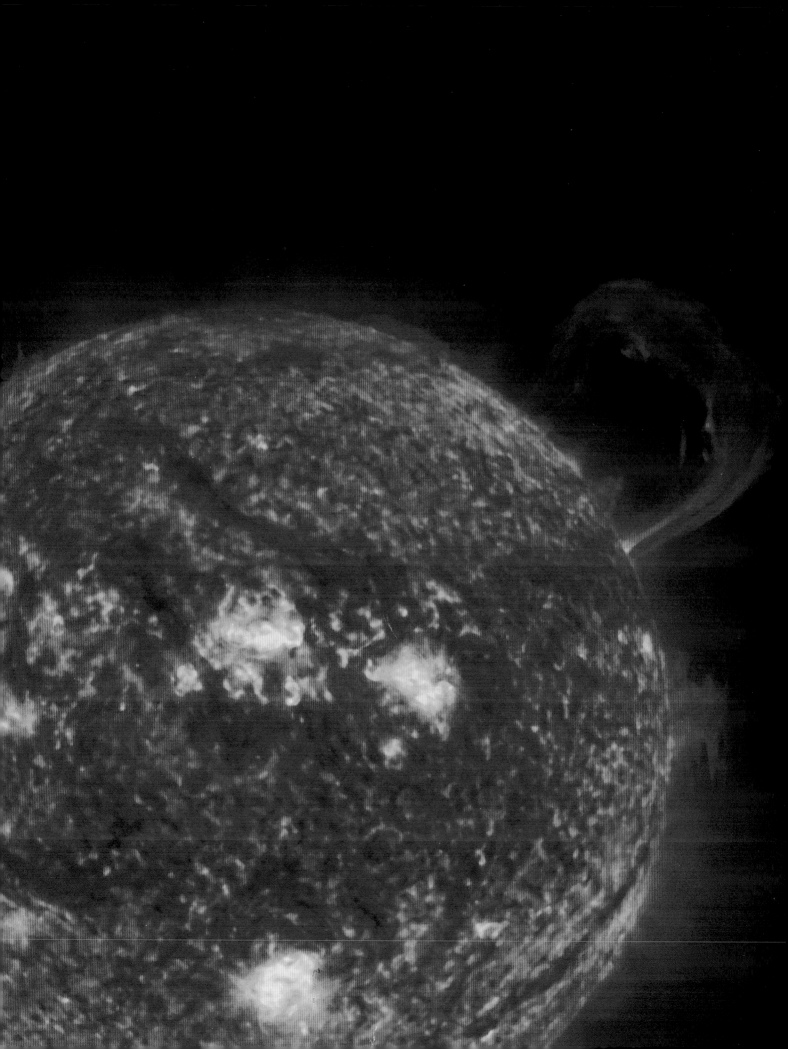

# THE
## SOLAR SYSTEM

Among the millions and millions of stars that form the Milky Way galaxy, there is a medium-sized one located in one of the galaxy's arms—the Sun. This star, together with the planets and other bodies that spin in orbits around it, make up the solar system, which formed about 4.6 billion years ago.

# THE SOLAR SYSTEM

The planets, satellites, asteroids, other rocky objects and the countless number of comets that circle the Sun comprise the Solar System; in total, it takes up a space of around 15 billion kilometers (9.3 billion miles) in diameter. The elliptical paths of the planets around the Sun are known as 'orbits.' This movement is attributable to the influence of the Sun's gravitational field in a dynamic balance. Today, thanks to advances in astronomy, it is known that there are more than fifty extrasolar planets and three multi-planetary systems.

## TYPES OF PLANET

Our Solar System is home to eight planets that orbit a single star: the Sun. The distance between the planets and the Sun determines the type of planet in question: the closest are rocky, smaller planets, whereas the most distant are gaseous, larger planets.

### THE CREATION OF THE PLANETS

The first proposals suggested that the planets were gradually formed by hot dust particles that joined together. Today, scientists believe that they were generated by the collision and merger of two large-sized bodies, known as 'planetesimals.'

**(1) Origins**
The remnants from when the Sun was created generate a disc of gas and dust, from which planetesimals form.

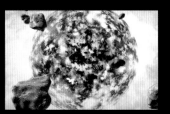

**(2) Collision**
When they collide together, the different sized planetesimals join together with other objects that have a greater mass.

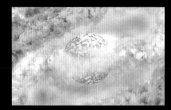

**(3) Heat**
These collisions generate a significant amount of heat inside the planets, depending on their distance from the Sun.

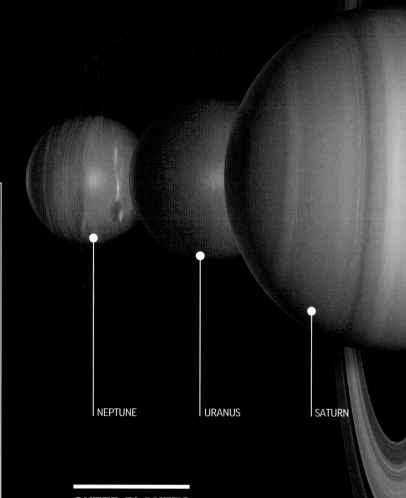

| NEPTUNE | URANUS | SATURN

## OUTER PLANETS

Planets located outside the asteroid belt. They are huge balls of gas with small, solid cores. Their temperatures are extremely low, due to the huge distance that separates them from the Sun; they are the only planets capable of sustaining planetary rings. The greatest in size is Jupiter, into which Earth would fit 1,300 times.

## SOLAR GRAVITY

The Sun's gravitational pull on the planets does not just keep them within the confines of the Solar System, it also influences the speed at which they rotate. Those closest to the Sun orbit more quickly than those further away.

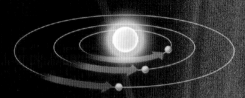

## INNER PLANETS

Planets located within the asteroid belt. They are solid bodies inside which inner geological phenomena take place, such as volcanism, capable of changing their surface. Almost all have a palpable atmosphere, although they differ in thickness; this plays a key role in the surface temperatures of each planet.

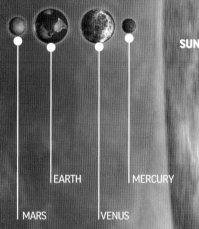

EARTH

MERCURY

MARS

VENUS

SUN

JUPITER

## ORBITS

In general, the planets orbit in one common plane called the ecliptic.

The rotation of most planets around their own axes is in counterclockwise direction. Venus and Uranus, however, revolve clockwise.

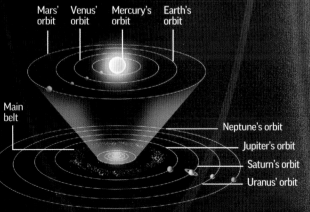

Mars' orbit
Venus' orbit
Mercury's orbit
Earth's orbit

Main belt

Neptune's orbit
Jupiter's orbit
Saturn's orbit
Uranus' orbit

## ASTEROID BELT

The boundary between the inner and outer planets is defined by an assembly of millions of rocky fragments of different sizes that form a ring known as the 'asteroid belt.' Its movement is seen to be influenced by the gravitational pull exercised by Jupiter.

# A VERY WARM HEART

**The Sun is an enormous ball of extremely dense and hot gases,** mainly hydrogen (90 percent) and helium (9 percent), with traces of elements such as carbon, nitrogen and oxygen, among others. To humankind, it is a vital source of light and heat; this energy is produced by the fusion of hydrogen atomic nuclei.

## CONVENTIONAL PLANETARY SYMBOL – SUN

### ESSENTIAL DATA

**Average distance from Earth**
150 million km (93 million miles)

**Diameter at the equator**
1.4 million km (864 million miles)

**Orbital speed**
220 km/sec (135 miles/sec)

**Mass (Earth = 1)**
332,900

**Gravity (Earth = 1)**
28

**Density**
0.255 g/cm3

**Average temperature**
5,500° C (9,932° F)

**Atmosphere**
Dense

**Moons**

## CONVECTIVE ZONE

This extends from the base of the photosphere to a depth of around 15 percent of the solar radius. Here, energy is transported outwards by (convective) currents of gas.

## RADIATIVE ZONE

Particles from the nucleus cross this zone. A photon can take half a million years to cross it.

# 8,000,000° C
(14,400,000° F)

## NUCLEAR FUSION

The extraordinary temperature of the nuclear core helps the hydrogen nuclei join. Under conditions of lower energy, they repel each other, but the conditions at the center of the Sun can overcome the repulsive forces, and nuclear fusion occurs. For every four hydrogen nuclei, a series of nuclear reactions produce one helium nucleus.

③
## HELIUM NUCLEUS

A group of two protons and a neutron collide with one another. A helium nucleus is formed, and a couple of protons are released.

Helium nucleus

Deuterium 1

Deuterium 2

Proton 1

Proton 2

① 
## NUCLEAR COLLISION

Two hydrogen nuclei (two protons) collide and join together. One transforms into a neutron and forms deuterium, releasing a neutrino, a positron and a significant amount of energy.

Positron

Proton

Neutron

Neutrino

Deuterium

Photon

② 
## PHOTONS

The deuterium formed collides with a proton. As a result of this collision, a gamma ray photon is released. The photon is full of energy and needs 30,000 years to reach the photosphere.

# SURFACE AND ATMOSPHERE

The visible portion of the Sun is a ball of light, comprised of boiling gases that emanate from its core. The flares of gas form plasma that crosses this layer. Then, they penetrate a vast layer of gases called the 'solar atmosphere'; here, two strata, the chromosphere and the corona, overlap. The energy generated by the Sun's core moves throughout the surface of the photosphere and the atmosphere for thousands of years.

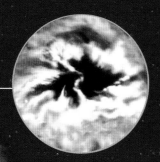

**SUNSPOTS**
These comprise areas of gas that are colder (4,000° C /7,232° F) than the photosphere (5,600° C/10,112° F); as a result, they are darker in appearance.

**NUCLEUS**
It occupies just 2 percent of the Sun's total volume; however, it is responsible for around half of its total mass. Given the intense pressures and temperature, thermonuclear fusions are generated here.

## 15,000,000° C
### (27,000,000° F)

**CHROMOSPHERE**
Above the photosphere, and with a much smaller density, this 5,000-km (3,107-mi) thick layer can be found. Its temperature ranges from 4,500° C to 500,000° C (8,132° F–900,032° F) depending on the distance from the core.

## 500,000° C
### (900,032° F)

**SPICULES**
These rising vertical jets of gas are attributable to the chromosphere. They often reach 10,000 km (6,214 mi) in height.

**MACROSPICULES**
These types of vertical jet are similar to spicules, but they often reach a height of 40,000 km (24,855 miles).

**PHOTOSPHERE**
This is the visible surface of the Sun, a boiling, thick tide of gases in a state of plasma. Density decreases while transparency increases in its outermost stratum. Thus, solar radiation escapes into extra-solar space in the form of light.

**CORONA**
Located above the chromosphere. It reaches millions of kilometers into space and extremely high temperatures.

## 1,000,000° C
### (1,800,000° F)

## 5,600° C
### (10,112° F)

**SOLAR PROMINENCES**
Clouds and layers of gas from the chromosphere that reach the corona. As a result of the activity of the magnetic fields to which they are subjected, they take on the form of an arc or a wave.

**SOLAR FLARES OR PROTUBERANCES**
These jets are released from the solar atmosphere and are capable of interfering with radio communication on Earth.

# MERCURY

**Mercury is the closest planet to the Sun, and as a result its average temperature can reach 167 degrees Celsius.** It moves at a high speed, orbiting the Sun every 88 days. It has practically no atmosphere and its surface is dry and harsh, plagued by craters caused by the impact of meteorites and numerous faults.

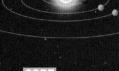

**CORE**
Dense, large and made of iron. Its diameter is believed to measure between 3,600 and 3,800 km (2,237 and 2,361 miles).

500 km
(311 miles)

3,600 km
(2,237 miles)

## A SCAR-COVERED SURFACE

On Mercury's surface, it is possible to find craters of different sizes, flatlands and hills. Recently, evidence of frozen water was found in the polar regions of Mercury. The polar ice may be located at the bottom of very deep craters, preventing the ice from interacting with sunlight.

**CALORIS BASIN**
Measuring 1,550 km (963 mi) in diameter, it is one of the largest craters in the entire Solar System. The largest is Utopia Planitia on Mars with a diameter of 3,300 km (2,050 mi).

The crater was submerged in lava.

When the projectile that formed the crater made impact, Mercury was still being formed: the expansive waves created hills and mountains.

**REMBRANDT**
The planet's second biggest basin (715 km [444 mi] in diameter).

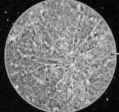

## SQUALID ATMOSPHERE

Mercury's atmosphere is almost non-existent: it consists of a very fine layer that is unable to protect the planet from the Sun or meteorites. As a result, temperatures during the day and at night vary enormously.

During the day, the Sun directly heats the surface.

At night, the surface quickly loses heat and the temperature drops.

**473° C**
(883° F)

**-183° C**
(-2970° F)

# COMPOSITION AND MAGNETIC FIELD

Like Earth, Mercury also has a magnetic field, although it is much weaker (around 1 percent). The magnetism is attributable to its huge core, comprised of solid iron. The mantle that encompasses the nucleus is made from a fine layer of liquid iron and sulphur.

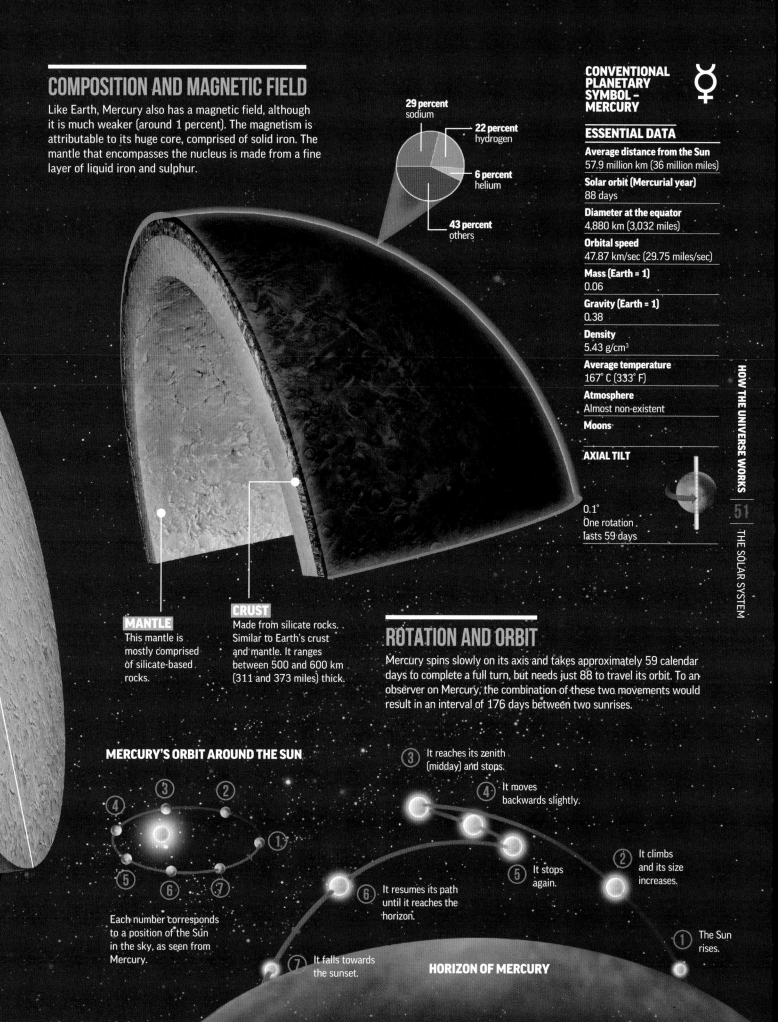

**29 percent**
sodium

**22 percent**
hydrogen

**6 percent**
helium

**43 percent**
others

## CONVENTIONAL PLANETARY SYMBOL – MERCURY

### ESSENTIAL DATA

**Average distance from the Sun**
57.9 million km (36 million miles)

**Solar orbit (Mercurial year)**
88 days

**Diameter at the equator**
4,880 km (3,032 miles)

**Orbital speed**
47.87 km/sec (29.75 miles/sec)

**Mass (Earth = 1)**
0.06

**Gravity (Earth = 1)**
0.38

**Density**
5.43 g/cm³

**Average temperature**
167° C (333° F)

**Atmosphere**
Almost non-existent

**Moons**

### AXIAL TILT

0.1°
One rotation
lasts 59 days

## MANTLE
This mantle is mostly comprised of silicate-based rocks.

## CRUST
Made from silicate rocks. Similar to Earth's crust and mantle. It ranges between 500 and 600 km (311 and 373 miles) thick.

## ROTATION AND ORBIT

Mercury spins slowly on its axis and takes approximately 59 calendar days to complete a full turn, but needs just 88 to travel its orbit. To an observer on Mercury, the combination of these two movements would result in an interval of 176 days between two sunrises.

### MERCURY'S ORBIT AROUND THE SUN

Each number corresponds to a position of the Sun in the sky, as seen from Mercury.

③ It reaches its zenith (midday) and stops.

④ It moves backwards slightly.

② It climbs and its size increases.

⑤ It stops again.

① The Sun rises.

⑥ It resumes its path until it reaches the horizon.

⑦ It falls towards the sunset.

**HORIZON OF MERCURY**

# VENUS

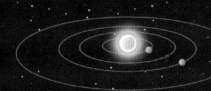

**Venus is the second closest planet to the Sun.** Similar in size to Earth, it has a volcanic surface and a hostile atmosphere, and is ruled by the effects of carbon dioxide. Four billion years ago, the atmospheres on Earth and Venus were similar; today, the atmosphere on Venus is 100 times more massive than on Earth. Its sulphuric acid and dust clouds are so thick and dense that it would be impossible to see the stars from the planet's surface.

## THE EFFECTS OF ITS THICK ATMOSPHERE

The predominant carbon dioxide content in Venus' atmosphere generates a greenhouse effect that elevates the planet's surface temperature to around 462° C (864° F). Consequently, Venus is hotter than Mercury, despite being further from the Sun, and with the fact that only 20 percent of sunlight reaches its surface (due to its dense atmosphere). Pressure on Venus is 90 times greater than the pressure on Earth.

# 8,000° C
## (14,432° F)
**THE CORE TEMPERATURE.**

## CORE
Believed to be similar to Earth's core, with metallic (iron and nickel) and silicate elements. It has no magnetic field, perhaps due to its slow rotating speed.

## MANTLE
Comprising molten rock, it is responsible for trapping solar radiation.

**CONVENTIONAL PLANETARY SYMBOL – VENUS** ♀

## ESSENTIAL DATA

**Average distance from the Sun**
108 million km (67 million miles)

**Solar orbit (Venusian year)**
224 days 17 hours

**Diameter at the equator**
12,100 km (7,519 miles)

**Orbital speed**
35.02 km/sec (21.76 miles/sec)

**Mass (Earth = 1)**
0.08

**Gravity (Earth = 1)**
0.9

**Density**
5.25 g/cm³

**Average temperature**
460° C (860° F)

**Atmosphere**
Very dense

**Moons**

**97 percent** carbon dioxide.

**3 percent** nitrogen and remnants of other gases.

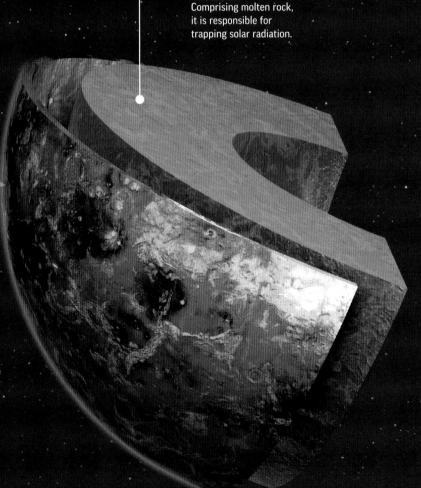

## ATMOSPHERE
Its glowing appearance is attributable to its thick, suffocating atmosphere, made up of carbon dioxide and sulphur clouds that reflect the Sun's light.

## AXIAL TILT

117°
One rotation takes 243 days

# 80 KM
## (49.7 MI)
**THE THICKNESS OF VENUS' ATMOSPHERE.**

## PHASES OF VENUS

While Venus orbits the Sun, its visibility from Earth is more or less apparent depending on its position compared with the Sun and our planet. That is to say, it has 'phases' like those of the Moon. It shines most during elongations (the angle between the Sun and the planet, as viewed from Earth), when it is furthest from the Sun in the sky. As a result, it can be seen after the Sun sets or before it rises.

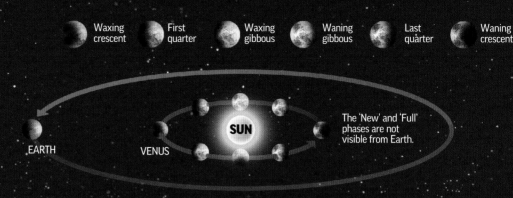

Waxing crescent · First quarter · Waxing gibbous · Waning gibbous · Last quarter · Waning crescent

SUN

EARTH

VENUS

The 'New' and 'Full' phases are not visible from Earth.

## SURFACE

Venus' surface has not remained the same since its creation. Its current surface is 500 million years old and this rocky surface is attributable to intense volcanic activity. The entire planet is characterized by wide plains, enormous rivers of lava and a number of mountains. The shine on the surface is attributable to its metallic compounds.

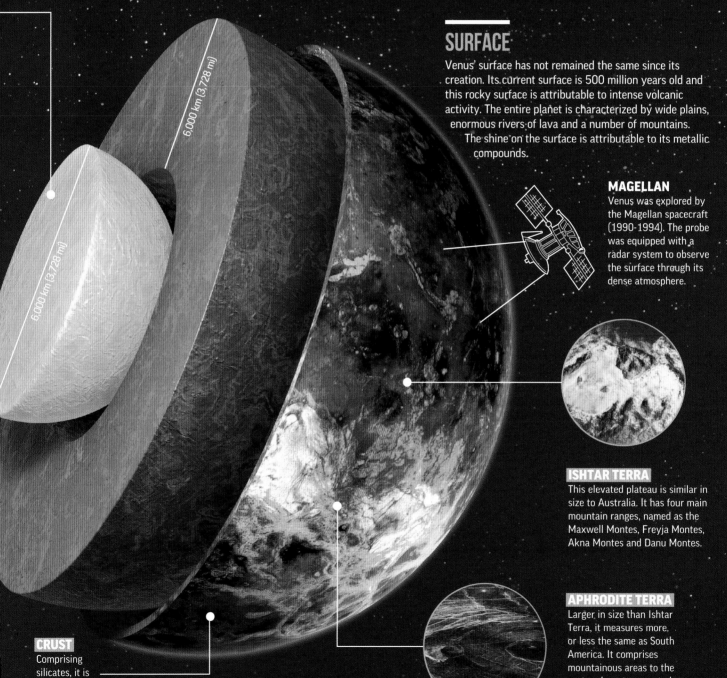

### MAGELLAN

Venus was explored by the Magellan spacecraft (1990-1994). The probe was equipped with a radar system to observe the surface through its dense atmosphere.

6,000 km (3,728 mi)

6,000 km (3,728 mi)

### ISHTAR TERRA

This elevated plateau is similar in size to Australia. It has four main mountain ranges, named as the Maxwell Montes, Freyja Montes, Akna Montes and Danu Montes.

### APHRODITE TERRA

Larger in size than Ishtar Terra, it measures more or less the same as South America. It comprises mountainous areas to the east and west, separated by lowlands.

### CRUST

Comprising silicates, it is thicker than Earth's crust.

# MARS

Known as the 'Red Planet' as its surface is covered in iron oxide, Mars' atmosphere is thin and not particularly dense; it is essentially comprised of carbon dioxide. Its orbital period, tilt axis and internal structure are all similar to those on Earth. Its poles house ice pockets and although no water can be seen on its surface, it is believed that the planet's water content was high during past times, and that there may be water in the sub-surface.

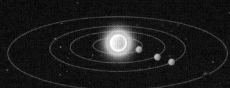

## MARTIAN ORBIT

Mars's orbit is more elliptical than Earth's; as a result, its distance from the Sun varies more. At its closest point, Mars receives 45 percent more solar radiation than at its furthest point. Surface temperatures vary between -140° C and 17° C (-220° F and 63° F).

**-140° C (-220° F)**
in winter

**17° C (63° F)**
in summer

SUN

Earth

Mars

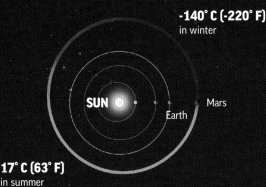

## MISSIONS TO MARS

After our own Moon, Mars has been a more attractive target for exploratory missions than any other object in the Solar System.

**MANTLE**
Molten rock, the density of which is greater than Earth's surface.

## MOONS

Mars has two moons, Phobos and Deimos, both of which are more dense than Mars and are pockmarked with craters. They are made up of carbon-rich rock. Deimos orbits Mars in 30.3 hours, whereas Phobos, which is closer to the red planet, does so in just 7.66 hours. Astronomers believe that the moons were asteroids attracted by Mars's gravity.

### DEIMOS

**Diameter**
15 km (9.3 mi)
**Distance from Mars**
23,540 km (14,627 mi)

### PHOBOS

**Diameter**
27 km (16.8 mi)
**Distance from Mars**
9,400 km (5,841 mi)

# 6,794 KM
## (4,222 MILES)

**THE DIAMETER OF MARS. IT MEASURES ALMOST HALF THE SIZE OF EARTH.**

## COMPOSITION

Mars, a rocky planet, has an iron-rich core. Mars is almost half the size of the Earth and has a similar period of rotation, as well as clearly evident clouds, winds, and other weather phenomena. Its thin atmosphere is made up of carbon dioxide, and its red color comes from its soil, which is rich in iron oxide.

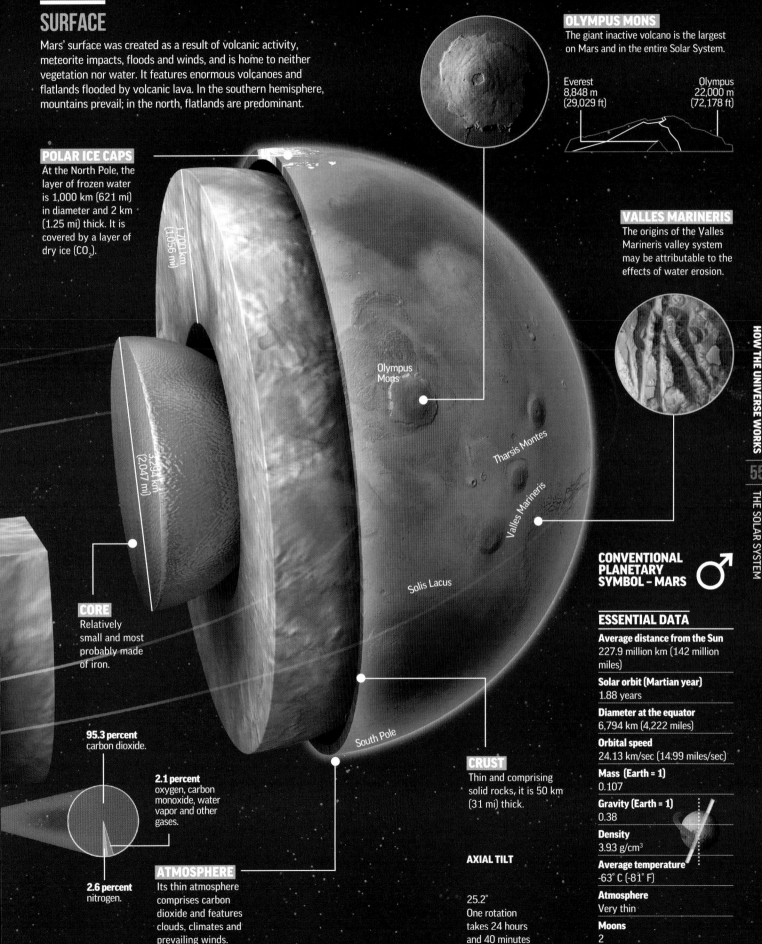

# SURFACE

Mars' surface was created as a result of volcanic activity, meteorite impacts, floods and winds, and is home to neither vegetation nor water. It features enormous volcanoes and flatlands flooded by volcanic lava. In the southern hemisphere, mountains prevail; in the north, flatlands are predominant.

## OLYMPUS MONS

The giant inactive volcano is the largest on Mars and in the entire Solar System.

Everest
8,848 m
(29,029 ft)

Olympus
22,000 m
(72,178 ft)

## POLAR ICE CAPS

At the North Pole, the layer of frozen water is 1,000 km (621 mi) in diameter and 2 km (1.25 mi) thick. It is covered by a layer of dry ice ($CO_2$).

1,700 km
(1,056 mi)

3,294 km
(2,047 mi)

## VALLES MARINERIS

The origins of the Valles Marineris valley system may be attributable to the effects of water erosion.

Olympus Mons

Tharsis Montes

Valles Marineris

Solis Lacus

South Pole

## CORE

Relatively small and most probably made of iron.

**95.3 percent** carbon dioxide.

**2.1 percent** oxygen, carbon monoxide, water vapor and other gases.

**2.6 percent** nitrogen.

## ATMOSPHERE

Its thin atmosphere comprises carbon dioxide and features clouds, climates and prevailing winds.

## CRUST

Thin and comprising solid rocks, it is 50 km (31 mi) thick.

## AXIAL TILT

25.2°
One rotation takes 24 hours and 40 minutes

## CONVENTIONAL PLANETARY SYMBOL – MARS ♂

## ESSENTIAL DATA

**Average distance from the Sun**
227.9 million km (142 million miles)

**Solar orbit (Martian year)**
1.88 years

**Diameter at the equator**
6,794 km (4,222 miles)

**Orbital speed**
24.13 km/sec (14.99 miles/sec)

**Mass (Earth = 1)**
0.107

**Gravity (Earth = 1)**
0.38

**Density**
3.93 g/cm³

**Average temperature**
-63° C (-81° F)

**Atmosphere**
Very thin

**Moons**
2

# JUPITER

**The largest planet in the Solar System.** Its diameter is eleven times greater than Earth's, its mass is 300 times greater and it spins at a speed of 40,000 km (24,855 miles) per hour. One of the most distinctive features of its atmosphere is the so-called 'Great Red Spot,' an enormous area of high pressure turbulence. The planet has several satellites and a fine ring of particles that orbit around it.

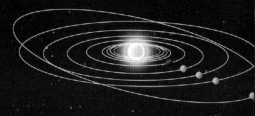

## COMPOSITION

Jupiter is a huge mass of hydrogen and helium, compressed in liquid form. Little is known about its core, and it has not been possible to measure its size; however, it is believed to be a metallic solid with a high density.

**CONVENTIONAL PLANETARY SYMBOL – JUPITER** ♃

### ESSENTIAL DATA

**Average distance from the Sun**
778 million km
(483 million mi)

**Solar orbit (Jovian year)**
11 years 312 days

**Diameter at the equator**
142,800 km (88,732 mi)

**Orbital speed**
13,07 km/sec (8.12 mi/sec)

**Mass (Earth = 1)**
318

**Gravity (Earth = 1)**
2.36

**Density**
1.33 g/cm³

**Average temperature**
-120° C (-184° F)

**Atmosphere**
Very dense

**Moons**
67

### AXIAL TILT

3.1°
One rotation takes 9 hours and 55 minutes

**CRUST**
It is 1,000 km (621 mi) thick.

37,700 Km (23,425 mi)

27,000 km (16,777 mi)

**CORE**

**INNER MANTLE**
Comprising metallic hydrogen, an element that can only be found at very high temperatures and pressures.

**OUTER MANTLE**
Comprising liquid hydrogen and helium. The outer mantle merges with the planet's atmosphere.

# 67 MOONS

HAVE BEEN DISCOVERED TO DATE. ADDITIONALLY, A FURTHER DOZEN 'TEMPORARY' MOONS HAVE BEEN CATALOGUED, PENDING CONFIRMATION OF THEIR NATURE AND ORBIT.

## GALILEAN MOONS

Of Jupiter's many moons, four can be seen from Earth with the use of binoculars. They are known as the Galilean moons, in honor of their discoverer, Galileo Galilei. It is believed that they house active volcanoes and Europa may be home to an ocean beneath its ice crust.

**GANYMEDE**
5,268 km
(3,273 mi)

**EUROPA**
3,200 km
(1,988 mi)

**IO**
3,643 km (2,264 mi)

**CALLISTO**
4,806 km (2,986 mi)

## RINGS
Formed by dust released by the planet's four inner moons.

# 26,000 KM
## (16,156 MI)

**IS THE LENGTH OF THE 'GREAT RED SPOT.'**

## WINDS

The planet's surface winds blow in opposite directions and in contiguous bands. The slight variations in their individual temperature and chemical composition are responsible for the planet's colored bands. The inclement environment – winds can exceed 600 km/h (373 mph) – are capable of causing storms, such as the Great Red Spot. It is believed to be mainly comprised of ammonia gas and ice clouds.

# 650,000,000 KM
## (403,891,275 MI)

**JUPITER'S MAGNETOSPHERE IS THE LARGEST OBJECT IN THE SOLAR SYSTEM. IT VARIES IN SIZE AND SHAPE DEPENDING ON ITS INTERACTION WITH SOLAR WINDS (MATTER RELEASED BY THE SUN EVERY SECOND).**

## THE MAGNETISM OF JUPITER
Jupiter's magnetic field is 20,000 times more intense than Earth's. The planet is surrounded by an enormous magnetic bubble – the magnetosphere. Its magnetotail extends beyond Saturn's orbit.

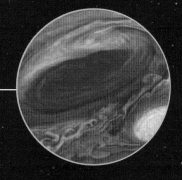

## ATMOSPHERE
It encompasses the inner liquid and solid core layers.

**89.8 percent** hydrogen.

**10.2 percent** helium with traces of methane and ammonia.

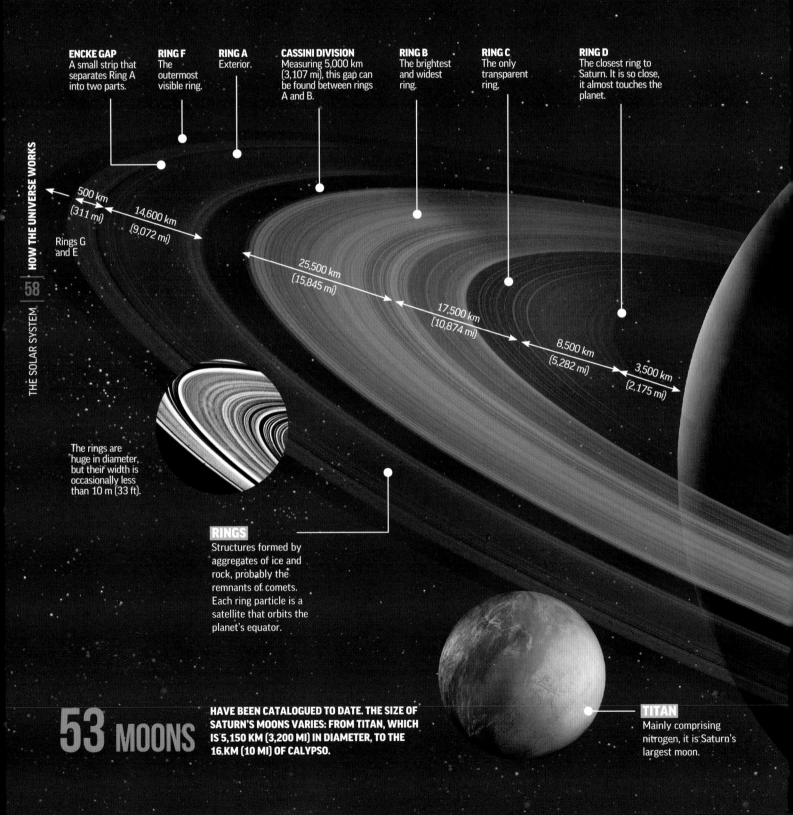

# SATURN

Just like Jupiter, Saturn is a huge ball of gas that encompasses a small solid core. To an onlooker, it would seem like just another yellow-tinted star; however, with the help of a telescope, its rings can be clearly distinguished. Ten times further from the Sun than Earth, it is the least dense of all the planets, and would even be able to float in the sea.

**ENCKE GAP**
A small strip that separates Ring A into two parts.

**RING F**
The outermost visible ring.

**RING A**
Exterior.

**CASSINI DIVISION**
Measuring 5,000 km (3,107 mi), this gap can be found between rings A and B.

**RING B**
The brightest and widest ring.

**RING C**
The only transparent ring.

**RING D**
The closest ring to Saturn. It is so close, it almost touches the planet.

500 km (311 mi)

14,600 km (9,072 mi)

Rings G and E

25,500 km (15,845 mi)

17,500 km (10,874 mi)

8,500 km (5,282 mi)

3,500 km (2,175 mi)

The rings are huge in diameter, but their width is occasionally less than 10 m (33 ft).

**RINGS**
Structures formed by aggregates of ice and rock, probably the remnants of comets. Each ring particle is a satellite that orbits the planet's equator.

# 53 MOONS
**HAVE BEEN CATALOGUED TO DATE. THE SIZE OF SATURN'S MOONS VARIES: FROM TITAN, WHICH IS 5,150 KM (3,200 MI) IN DIAMETER, TO THE 16 KM (10 MI) OF CALYPSO.**

**TITAN**
Mainly comprising nitrogen, it is Saturn's largest moon.

# SURFACE

Saturn has a surface of clouds, which form bands attributable to the rotation of the planet on its axis. Saturn's clouds are calmer and less colourful than those of Jupiter. Temperatures in the highest (white) clouds reach -140° C (-220° F) and a layer of fog extends over them.

Fog

White clouds

Deep, orange clouds

Blue clouds

## ATMOSPHERE

Mainly comprised of hydrogen and helium. The remainder is made up of sulphurs (responsible for its yellowy tones), methane and other gases.

1 percent sulphurs and other gases.

97 percent hydrogen.

2 percent helium.

## ESSENTIAL DATA

**Average distance from the Sun**
1,427 million km (887 million miles)

**Solar orbit (Saturnian year)**
29 years 154 days

**Diameter at the equator**
120,600 km (74,937 miles)

**Orbital speed**
9.66 km/sec (6 miles/sec)

**Mass (Earth = 1)**
95

**Gravity (Earth = 1)**
0.92

**Density**
0.69 g/cm³

**Average temperature**
-125° C (-193° F)

**Atmosphere**
Very dense

**Named Moons**
53

## AXIAL TILT

26.7°
One rotation takes 10 hours and 39 minutes

## WINDS

Wind speeds of up to 360 km/h (224 mph) can be reached at the equator. The planet can experience torrid storms.

## MANTLE

The planet is externally covered by a mantle of liquid hydrogen and helium that extends into the planet's gaseous atmosphere.

30,000 km (18,641 mi)

14,000 km (8,699 mi)

32,000 km (19,884 mi)

## HYDROGEN LAYER

Liquid hydrogen encompasses the outer core.

## CORE

Comprising rock and metallic elements such as silicates and iron. On the inside, it is similar to Jupiter.

# 12,000° C
(21,632° F)

**THE CORE TEMPERATURE.**

## OUTER CORE

Water, methane and ammonia encompass the hot rocky core.

# URANUS

At first sight, Uranus seems like just another star at the furthest limits of the naked eye's reach. It is almost four times larger than Earth and is unique in that its rotation axis is tilted to almost 98 degrees relative to its its orbital plain, meaning one of its poles is always facing the Sun. Uranus's orbit is so large that the planet takes 84 years to orbit the Sun just once.

## MAGNETIC FIELD

Uranus's magnetic field is 50 times greater than Earth's and is tilted 60 degrees compared to its rotation axis. On Uranus, magnetism is generated by the mantle and not the core.

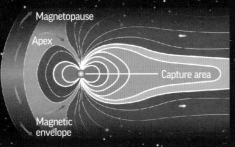

Magnetopause

Apex

Capture area

Magnetic envelope

Some scientists have suggested that Uranus's strange magnetic field may be attributable to the fact that there is no convection at its core due to cooling, or that it is magnetically inverted.

## CONVENTIONAL PLANETARY SYMBOL – URANUS

### ESSENTIAL DATA

**Average distance from the Sun**
2,870 million km (1,783 million miles)

**Solar orbit (Uranian year)**
84 years 36 days

**Diameter at the equator**
51,800 km (32,187 miles)

**Orbital speed**
6.82 km/sec (4.24 miles/sec)

**Mass (Earth = 1)**
14.5

**Gravity (Earth = 1)**
0.89

**Density**
1.32 g/cm³

**Average temperature**
-210° C (-346° F)

**Atmosphere**
Not dense

**Moons**
27

### AXIAL TILT

97.9°
One rotation takes 17 hours and 14 minutes

## 10,000° C
(18,032° F)

### THE CORE TEMPERATURE.

**CORE**
Comprising siliceous rocks and ice.

**MANTLE 1**
Comprising water, ice, methane gas, ammonia and ions.

**MANTLE 2**
Around the mantle, there may be another layer of liquid molecular hydrogen and liquid helium with a small amount of methane.

**ATMOSPHERE**
Comprising hydrogen, methane, helium and small amounts of acetylene and other hydrocarbons.

## -210° C
(-346° F)

### AVERAGE TEMPERATURE.

10,000 km (6,214 mi)

17,000 km (10,563 mi)

10,000 km (6,214 mi)

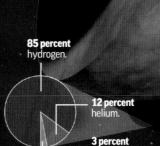

**85 percent** hydrogen.

**12 percent** helium.

**3 percent** methane.

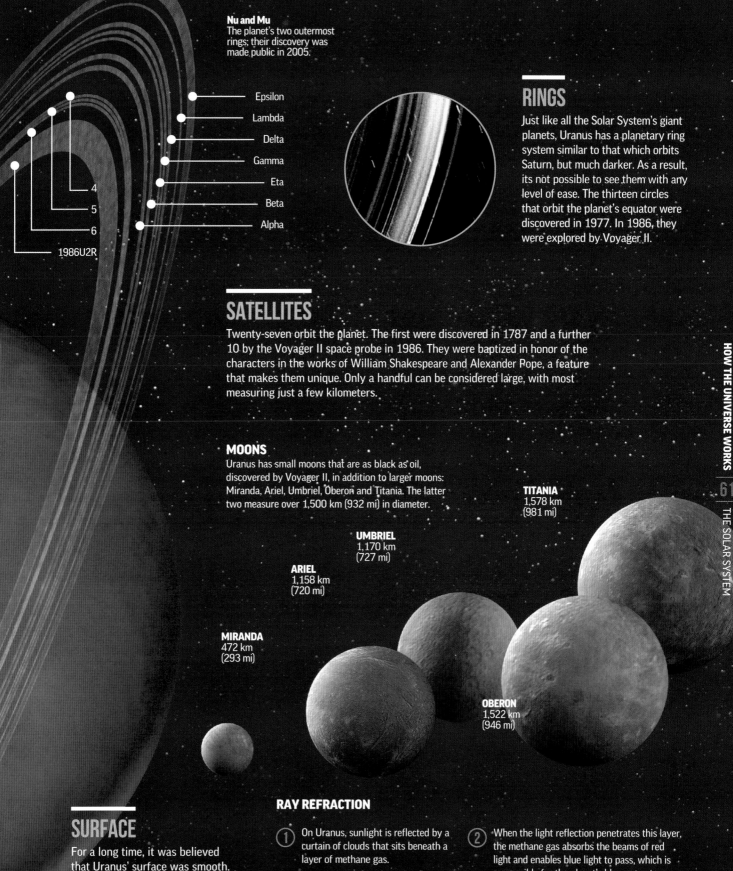

**Nu and Mu**
The planet's two outermost rings; their discovery was made public in 2005.

Epsilon
Lambda
Delta
Gamma
Eta
Beta
Alpha

4
5
6
1986U2R

## RINGS

Just like all the Solar System's giant planets, Uranus has a planetary ring system similar to that which orbits Saturn, but much darker. As a result, its not possible to see them with any level of ease. The thirteen circles that orbit the planet's equator were discovered in 1977. In 1986, they were explored by Voyager II.

## SATELLITES

Twenty-seven orbit the planet. The first were discovered in 1787 and a further 10 by the Voyager II space probe in 1986. They were baptized in honor of the characters in the works of William Shakespeare and Alexander Pope, a feature that makes them unique. Only a handful can be considered large, with most measuring just a few kilometers.

## MOONS

Uranus has small moons that are as black as oil, discovered by Voyager II, in addition to larger moons: Miranda, Ariel, Umbriel, Oberon and Titania. The latter two measure over 1,500 km (932 mi) in diameter.

**TITANIA**
1,578 km
(981 mi)

**UMBRIEL**
1,170 km
(727 mi)

**ARIEL**
1,158 km
(720 mi)

**MIRANDA**
472 km
(293 mi)

**OBERON**
1,522 km
(946 mi)

## SURFACE

For a long time, it was believed that Uranus' surface was smooth. However, the Hubble telescope showed that it is a dynamic planet with the brightest clouds in the Solar System, and with a weak planetary ring system that oscillates like an unbalanced wheel.

## RAY REFRACTION

① On Uranus, sunlight is reflected by a curtain of clouds that sits beneath a layer of methane gas.

② When the light reflection penetrates this layer, the methane gas absorbs the beams of red light and enables blue light to pass, which is responsible for the planet's blue-green tone.

Atmosphere
Rays of sunlight
Uranus

Atmosphere
Rays of sunlight
Uranus

# NEPTUNE

The Solar System's outermost gas planet is 30 times further from the Sun than Earth, and looks like an extraordinary blue ball. This effect is attributed to the presence of methane in the outermost part of its atmosphere. Its moons, rings and incredible clouds all stand out, and its similarity to Uranus is also discernible. To scientists, Neptune is particularly special: its existence was proposed based on mathematical calculations and predictions.

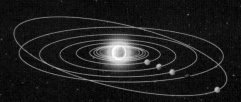

## MOONS

Neptune has 13 natural satellites. Triton and Nereid, those furthest from the planet, were the first to be seen from Earth using a telescope. The remaining 11 were observed from space by US spacecraft Voyager II. All their names correspond to gods of the sea from Greek mythology.

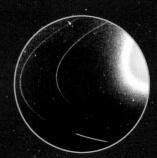

### COMPOSITION

Its rings are dark, although their composition is not known; it is also believed that they are not stable. For example, Liberty is the outermost part of the ring and it may completely vanish by the end of this century.

### TRITON

Measuring 2,706 km (1,681 mi) in diameter, it is Neptune's largest moon. It orbits the planet in the opposite direction to all its other moons, and its surface is marked with dark grooves, formed by the dust that it precipitates following the eruptions of its geysers and volcanoes.

## RINGS

From Earth, they look like arcs. However, we now know that they are rings of dust that shine, reflecting rays of sunlight. Their names honor the first scientists who studied the planet.

GALLE

URBAIN
LE VERRIER

LASSELL

ARAGO

### ADAMS

Located 63,000 km (39,146 mi) from the planet's core. It is a formation of three intertwining rings named Liberty, Fraternity and Equality.

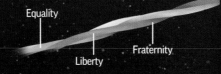

Equality

Liberty

Fraternity

## -235° C
## (-391° F)

**THE TEMPERATURE ON TRITON, ONE OF THE SOLAR SYSTEM'S COLDEST BODIES.**

# SURFACE

White methane clouds encompass the planet. The winds circulate from east to west, in the opposite direction to the planet's rotation, reaching speeds of 2,000 km/h (1,243 mph).

Ascending winds

Descending winds

## GREAT DARK SPOT

A giant storm, the size of Earth, similar to the Great Red Spot on Jupiter, stood out against Neptune's surface. It was first seen in 1989 and broke up in 1994.

7,200 km (4,474 mi)

14,000 km (8,699 mi)

6,000 km (3,728 mi)

## CORE

The typical core of gas planets is repeated on Neptune – a rocky sphere that turns molten towards the surface.

# STRUCTURE

It has a rocky silicate core, covered by a layer of frozen water, ammonia, hydrogen and methane, known as the 'mantle.' The core and mantle occupy two-thirds of Neptune's interior. The final third is the thick, dense atmosphere which consists of a mix of hot gases, comprising hydrogen, helium, water and methane.

## CONVENTIONAL PLANETARY SYMBOL – NEPTUNE

### ESSENTIAL DATA

**Average distance from the Sun**
4,500 million km (2,796 million mi)

**Solar orbit (Neptunian year)**
164 years 264 days

**Diameter at the equator**
49,500 km (30,758 mi)

**Orbital speed**
5.48 km/sec (3.41 miles/sec)

**Mass (Earth = 1)**
17.2

**Gravity (Earth = 1)**
1.12

**Density**
1.64 g/cm$^3$

**Average temperature**
-200° C (-382° F)

**Atmosphere**
Dense

**Moons**
13

**AXIAL TILT**

28.3°
One rotation takes 16 hours and 36 minutes

## MANTLE 1

The component materials of this layer convert from a solid to a gaseous state.

## MANTLE 2

Containing a higher level of gaseous material than solid material.

## ATMOSPHERE

The gases that make up the atmosphere are concentrated in similar bands as those found on other gas giants. They form a cloud system that is as active, or even more active, than the system on Jupiter.

**10.2 percent** helium.

**89.8 percent** hydrogen.

# PLUTO

**Until 2006, Pluto was considered the ninth planet in the Solar System.**
That year however, the International Astronomical Union (IAU) decided
to designate it a 'dwarf planet.' Little is known about this tiny body in the Solar
System. However, some of its characteristics make it particularly special: its unique
orbit, its axial tilt and the fact that it is an object belonging to the Kuiper Belt.

## CHARON

Charon is Pluto's largest satellite. Incredibly, the diameter of
Pluto's biggest moon is almost half that of the planet itself. Its
surface appears to be covered in ice, unlike Pluto, the surface of
which comprises frozen nitrogen, methane and carbon dioxide.
One theory is that Charon was formed from ice that was torn
from Pluto following a collision with another object.

Pluto

Rotating
axis

Charon

### SYNCHRONIZED ORBITS

It is often considered that Pluto and Charon form a double planetary system.
The rotation between the two bodies is unique: their sightline is never broken,
and it appears like they are united by an invisible bar. They are synchronized
to such an extent that Charon can only, and permanently, be seen from one
side of Pluto, while from the other side, the moon is never seen.

# 1,172 KM
## (728 MI)

**THE DIAMETER OF CHARON,
HALF THAT OF PLUTO.**

## COMPOSITION

Based on different calculations, it
has been deduced that 75 percent
of Pluto consists of rocks mixed
with ice. Scientists have concluded
that Pluto forms part of the Kuiper
Belt, made up of material left
over from other planets. Large
blocks of frozen nitrogen aside,
it also features simple molecules
containing hydrogen and oxygen –
the indispensable sources of life.

## OTHER MOONS

Besides Charon, discovered
in 1978, Pluto has four other
moons: Nix and Hydra, discovered
in 2005 by the Hubble telescope,
and two further moons (P4
and P5) that remain unnamed,
discovered in 2011 and 2012.

### DENSITY

The density of Charon is 1.7 g/cm3 (0.98 oz/
cu in); it is therefore assumed that rocks do
not represent a large part of its composition.

# SURFACE

Little is known about Pluto. However, the Hubble telescope has shown that its surface is covered by a combination of frozen nitrogen and methane. The presence of methane in its solid state would suggest that the surface temperature is less than -203° C (-333° F). However, this depends on the point at which this 'dwarf planet' is in orbit, given its distance from the Sun can range between 30 and 50 astronomical units.

## ESSENTIAL DATA

**Average distance from the Sun**
5,900 million km (3,666 million miles)

**Solar orbit (Plutonian year)**
247.9 years

**Diameter at the equator**
2,247 km (1,396 miles)

**Orbital speed**
4.75 km/sec (2.95 miles/sec)

**Mass (Earth = 1)**
0.002

**Gravity (Earth = 1)**
0.062

**Density**
2.05 g/cm³

**Average temperature**
-230° C (-382° F)

**Atmosphere**
Very thin

**Moons**
5

**AXIAL TILT**

122°
One rotation takes 153 hours

### MANTLE
The mantle, comprising a layer of frozen water, encompasses the planet's core.

### CORE
Comprising a blend of iron, nickel and rock, its exact diameter remains unknown.

### CRUST
Made up of frozen methane and water at the surface; it has been calculated that it may be between 100 and 200 km (62 and 124 mi) thick.

**2 percent** methane, with some traces of carbon monoxide.

**98 percent** nitrogen.

### ATMOSPHERE
Pluto has a very fine atmosphere that is frozen and falls to the planet's surface as its orbit gets further away from the Sun.

## A UNIQUE ORBIT

Pluto's orbit is considerably elliptical and inclined (17° with regards to other planets). It is located between 4 and 7 billion km (2.5 and 4.3 billion mi) from the Sun. For 20 years of the planet's 248-year long orbit, Pluto is closer to the Sun than Neptune. Although it would appear that Pluto's orbit crosses paths with Neptune, it would be impossible for the planets to collide.

# DISTANT WORLDS

**Way beyond Neptune, there are a group of frozen bodies that are smaller in size than the Moon.** There are more than 100,000 objects that form the so-called Kuiper Belt, the frozen boundary of our Solar System. The belt is the main repository for 'short periodic comets,' those that appear with regular frequency. Other comets in turn originate in the huge sphere known as the Oort cloud, which encompasses the entire Solar System.

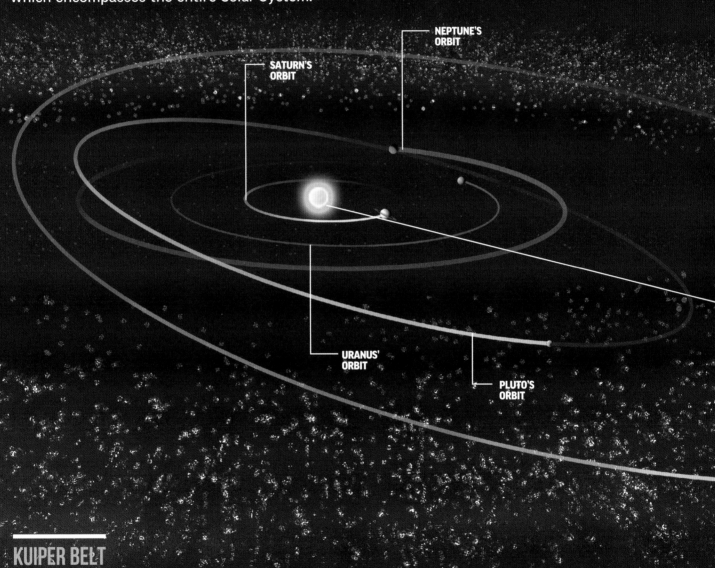

NEPTUNE'S ORBIT

SATURN'S ORBIT

URANUS' ORBIT

PLUTO'S ORBIT

## KUIPER BELT

Close to Neptune, in the Kuiper Belt, there are small frozen worlds that are similar to the Solar System's eighth planet, but which are much smaller in size. The belt comprises 100,000 chunks of ice and rock (including Pluto) spread out in the form of a ring, almost a thousand of which have been catalogued. The Kuiper Belt received its name from astronomer Gerard Kuiper, who predicted its existence in 1951, 40 years before it was seen for the first time.

# COMPARATIVE SIZES

The discovery of Quaoar in 2002 offered scientists a much sought-after link between the origin of the Solar System and the Kuiper Belt. Quaoar's orbit, which is almost circular, demonstrated that objects in the Kuiper Belt orbit around the Sun. Since 2006 and to date these objects have been considered by the International Astronomical Union as part of the 'dwarf planet' category to which Pluto belongs.

**QUAOAR**
Diameter:
around 1,300 km
(808 mi).

**SEDNA**
Diameter:
around 1,600 km
(994 mi).

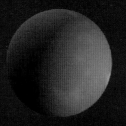

**PLUTO**
Diameter:
2,247 km
(1,396 mi).

**ERIS**
Diameter:
around 2,326 km
(1,445 mi).

# 35,000
**OBJECTS IN THE KUIPER BELT ARE ESTIMATED TO MEASURE OVER 100 KM (62 MI) IN DIAMETER.**

ERIS

## THE OUTERMOST DWARF PLANET
Eris is a dwarf planet located 95.7 astronomical units (14.3 billion km/9 billion mi) from the Sun, making it the most distant object observed in the Solar System. It appears that this planet follows an eccentric orbit (very elongated), which takes 557 years to complete. It has one moon, Dysnomia.

# ASTEROIDS AND METEORITES

Since the time the Solar System began to form, the fusion, collision and break-up of different materials has played an essential role in the formation of the planets. These 'small' rocks are a remnant of this process. They are witnesses that provide data to help the understanding of the extraordinary phenomena that began 4.6 billion years ago. On Earth, these objects are associated with episodes that would later influence evolutionary processes.

## THE NATURE OF METEORITES

One of the main goals in the study of meteorites is to determine their make-up. They contain both extraterrestrial gases and solids. Scientific tests have made it possible to confirm that, in some cases, the objects came from the Moon or Mars. However, for the most part, meteorites are associated with asteroids.

### HOW A METEORITE MAKES IMPACT

When penetrating Earth's atmosphere, they do not completely vaporize – on reaching the Earth's surface they leave a footprint called a 'crater.' Furthermore, they contribute exotic rock material to Earth's surface, such as large amounts of iridium, an element that is scarce on Earth, but common to the composition of meteorites.

## TYPES OF METEORITES

### AEROLITES
Notable for their olivine and pyroxene content. This category subdivides into chondrites and non-chondrites.

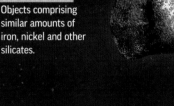

### IRON METEORITES
Abundant in iron-nickel compounds. Generated during the break-up of asteroids.

### SIDEROLITES
Objects comprising similar amounts of iron, nickel and other silicates.

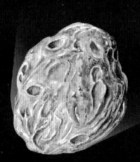

### ① EXPLOSION
Friction with the air increases the temperature of the meteorite. Thus, it starts to ignite.

## 12 KM/SEC
## (7.5 MI/SEC)
**THE IMPACT SPEED OF A METEORITE ON EARTH.**

### ② DIVISION
This fragmentation leads to a visual effect, known on Earth as 'shooting stars.'

### ③ IMPACT
On impact, it is compressed and carves out a hole in Earth's surface, creating a crater.

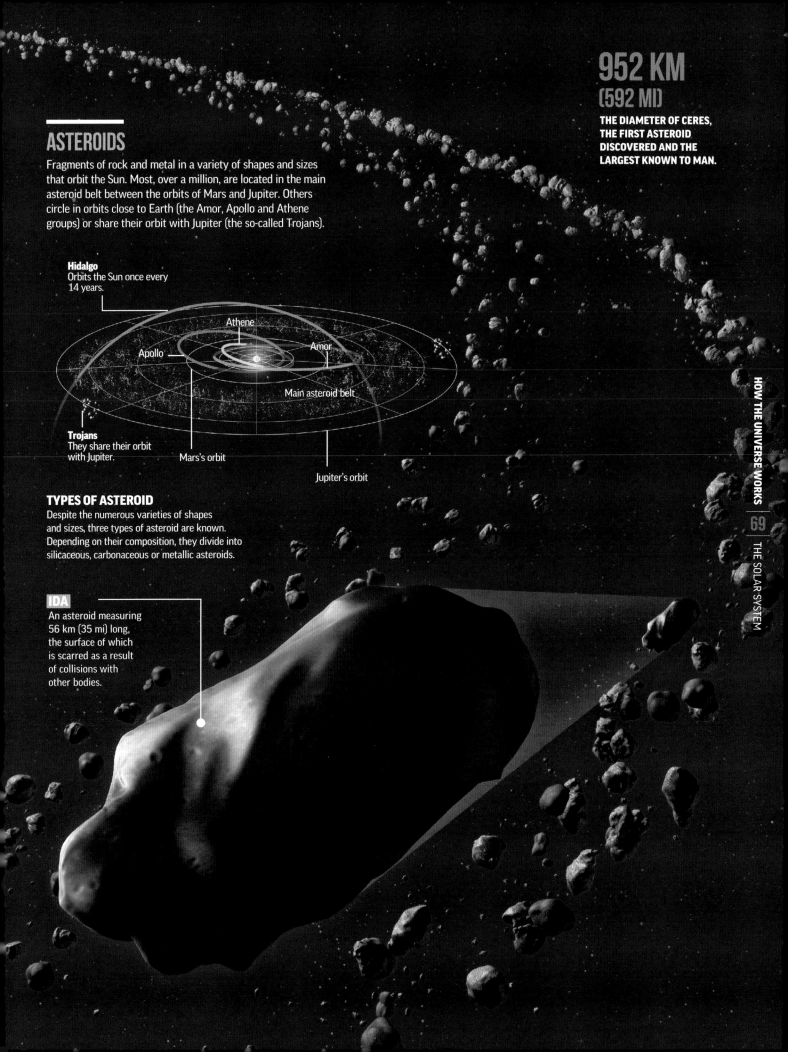

## ASTEROIDS

Fragments of rock and metal in a variety of shapes and sizes that orbit the Sun. Most, over a million, are located in the main asteroid belt between the orbits of Mars and Jupiter. Others circle in orbits close to Earth (the Amor, Apollo and Athene groups) or share their orbit with Jupiter (the so-called Trojans).

**952 KM**
**(592 MI)**

THE DIAMETER OF CERES, THE FIRST ASTEROID DISCOVERED AND THE LARGEST KNOWN TO MAN.

**Hidalgo**
Orbits the Sun once every 14 years.

Athene

Apollo

Amor

Main asteroid belt

**Trojans**
They share their orbit with Jupiter.

Mars's orbit

Jupiter's orbit

## TYPES OF ASTEROID

Despite the numerous varieties of shapes and sizes, three types of asteroid are known. Depending on their composition, they divide into silicaceous, carbonaceous or metallic asteroids.

**IDA**
An asteroid measuring 56 km (35 mi) long, the surface of which is scarred as a result of collisions with other bodies.

# COMETS

Comets are small, irregularly shaped objects measuring just a few kilometers in diameter that are usually frozen and dark in color. They are made of dust, rock, gases and organic molecules rich in carbon, and can be found orbiting in the Kuiper Belt or the so-called Oort cloud. However, many deviate towards the inner part of the Solar System, assuming new paths. When they warm up, their ice formations sublimate, forming their heads and long tails of gas and dust.

## TYPES OF COMET

Short-period comets are those that orbit the Sun in less than 200 years. Long-period comets, with an orbital period of over 200 years, have orbits dozens, or hundreds, of times greater than Pluto's.

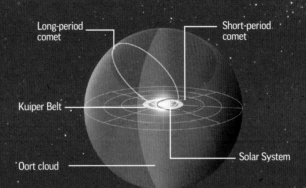

Long-period comet

Short-period comet

Kuiper Belt

Oort cloud

Solar System

## DEEP IMPACT SPACE MISSION

On 12 January 2005, as part of the Discovery Program, the US Space Agency launched Deep Impact. This spacecraft was designed to launch a projectile that impacted against the comet 9P/Tempel 1 to obtain samples to be studied on Earth.

① **Launch of the probe**
Deep Impact launches a copper projectile weighing 350 kg (772 lb) designed to collide with the comet.

② **In position**
using infrared cameras and spectrometers, the craft follows the comet to analyze the impact at its nucleus.

The projectile searches for the point of impact.

**SOLAR WIND**

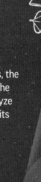

③ **Impact with the comet**
Took place on 4 July 2005. The projectile generated a crater the size of a football field and carved out a hole seven stories deep.

**COMA**
Covering the nucleus. Comprising gases and dust released by the nucleus.

**CORE**
Frozen water, methane, $CO_2$ and ammonia.

**HEART**
The innermost part contains powdered silicates.

**HEAD**
Comprising the nucleus and the coma. The front part is known as the point of impact.

# 36,000 KM/H
## (22,369 MPH)

**THE IMPACT SPEED AGAINST THE COMET.**

# 76 YEARS

**THE TIME IT TAKES HALLEY'S COMET TO COMPLETE ITS ORBIT.**

TAIL

HEAD

## ENVELOPE
Layers of hydrogen that are capable of forming a third tail.

## TAIL OF DUST
Suspended dust particles form a wake that reflects sunlight, making the comet's luminous tail visible.

## TAIL OF IONS
The tail of suspended gases generates a low intensity, luminous blue-colored area. Gas molecules lose an electron and acquire an electrical charge.

Close to the Sun, the tails get longer.

Moving away from the Sun, the tails disappear.

## FORMATION OF THE TAIL AND HEAD
Due to the effects of solar winds, when the comet gets closer to the Sun, the gases released travel further away. Meanwhile, the dust particles tend to form a wake that is curved, as it is less sensitive to the pressure of solar winds. As the comet travels further away from the confines of the Solar System, the tails merge back together, disappearing when the nucleus cools down and stops releasing gas.

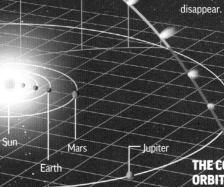

Sun

Earth

Mars

Jupiter

**THE COMET'S ORBIT**

# EXTRASOLAR PLANETS

Although the existence of exoplanets—planets that orbit a star other than a sun—was suspected for many centuries, it was not confirmed until the 1990s. The construction of powerful terrestrial and space telescopes like the Kepler space observatory—best known as the "catcher of the planets"—has accelerated the finding of extrasolar planets at a rapid pace.

## GAS OR ICE GIANTS

Since the first discovery in 1995 by Swiss astronomers Michel Mayor and Didier Queloz, there have been nearly 3,500 exoplanets discovered. The majority of which conform to a similar pattern: a massive gas or ice giant with an orbit close to their host star and with accordingly short orbit periods. This is why they are easily detected. Nevertheless, technological advancements in recent years are giving way to even more discoveries of Earthlike planets.

## 3.458
**EXOPLANETS HAVE BEEN CONFIRMED.**

## 2.581
**STELLAR SYSTEMS HAVE BEEN DISCOVERED.**

**Iota Draconis b.**
The first exoplanet orbiting a giant star is discovered. (Thirteen times larger than our sun.)

## CHRONOLOGY OF DISCOVERIES

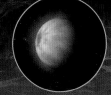

**First exoplanets discovered.**
A planetary system is detected orbiting the PSR B1257+12 Pulsar.

**51 Pegasi b.**
The first exoplanet discovered is half the size of Jupiter and orbits a Sun-like star.

**Upsilon Andromedae.**
The first binary planetary system is detected around another star. It is comprised of three Jupiter-like planets.

**HD 209458 b.**
First exoplanet is observed transiting its star.

**Atmosphere.**
Thanks to the Hubble Space Telescope, the extrasolar planet HD 209458 b's atmosphere is analyzed.

**HD 20185 b.**
The first time an exoplanet is discovered within the "habitable zone" of a star.

| 1990 | 1991 | 1992 | 1990 | 1991 | 1992 | 1993 | 1994 | 1995 | 1996 | 1997 | 1998 | 1999 | 2000 | 2001 | 2002 |

## PROXIMA B

**Artist's rendering of exoplanet that orbits the dwarf red star Proxima Centauri, the closest to the sun.**

# METHODS OF DETECTION

The light that an extrasolar planet emits is minimal and drowned out by the powerful glow of their star, which makes observing them directly difficult and complicated. We use methods of indirect detection to calculate the presence and typology of some planets based on the movement variation or brightness of the star. Transit Photometry (Transit) detects 79.5% of distant planets found by measuring the minute dimming of a star as an orbiting planet passes between it and the Earth. Doppler Spectroscopy (Radial-Velocity, or colloquially, the wobble method) is responsible for 17.9% extrasolar planets found. These two methods are the most effective.

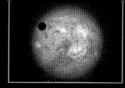

**TRANSIT**

The path of a plan its star and the po observation provo detectable variatic glare of the star.

**RADIAL VELO**

The orbiting plane slight gravitationa star, varying its m and wavelengths c spectral lines.

## WANDERING BODIES

Also known as interstellar planets, they are not subject to the gravitational force of any star because they were expelled from the stellar system in which they were formed.

**Fomalhaut b.**
The first exoplanet photographed directly by the Hubble Space Telescope.

**Kepler-452b.**
Another Earthlike exoplanet is discovered but this time orbiting a Sun-like star.

**Spectroscope.**
First spectral measurement of HD 209458 b and HD 189733-b -image atmosphere. It is considered the mechanism for finding extrasolar life.

**Emits its own light.**
The Spitzer telescope first observed infrared light emitted from two exoplanets: HD 209458 b and TrES-1.

**Kepler-10b.**
The Kepler mission discovers the first rocky exoplanet, the smallest planet outside the solar system.

**Kepler-186f.**
It is the first time an Earth-like planet orbiting in the "habitable zone" of a star is found.

**TRAPPIST-1.**
Four new planets have been detected around this star of Earth-like composition and size. There are now seven in total.

2003    2004    2005    2006    2007    2008    2009    2010    2011    2012    2013    2014    2015    2016    20

# LIKE THE EARTH

**Space agencies started searching for Earth's "twin" as soon as the existence of other planets outside the solar system were discovered.** In recent years there have been extraordinary advances, with more terrestrial planets detected that may have the minimal conditions necessary for the existence of water and, consequently, life.

## EARTH SIMILARITY INDEX (ESI)

A scale created to estimate how habitable an exoplanet is as it relates to Earth. ESI assigns a number from zero to one, with an ESI of one being identical to Earth. This number is calculated from the exoplanet's radius, density, surface temperature and escape velocity. Exoplanets with an ESI of 0.8 or greater are likely to contain a rocky composition with similar temperatures to Earth. Kepler-438b has the highest ESI of any exoplanet known, with an ESI of 0.88.

## 352

**NUMBER OF EXOPLANETS DISCOVERED CONSIDERED TERRESTRIAL, WITH CHARACTERISTICS SIMILAR TO EARTH.**

### KEPLER-442B
Another exoplanet that has an elevated ESI of 0.84 and is 1,105.5 light years from Earth.

## THE "HABITABLE ZONE"

Having an elevated ESI does not guarantee a habitable environment hospitable to life. Another necessary factor is that the planet must be within the "habitable zone" of its star. It is a range of orbits around a star within which a planetary surface can support liquid water given sufficient atmospheric pressure. Other potential factors that influence the habitability of a planet are its eccentricities, including its orbital properties and atmospheric conditions. In the illustration, the "habitable zone" of the Solar System is compared to the Kepler-186 and Kepler-452 star systems.

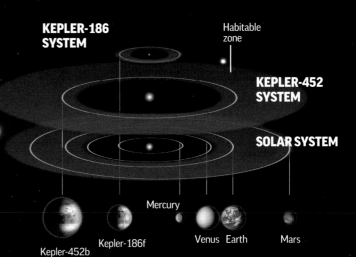

**KEPLER-186 SYSTEM**

Habitable zone

**KEPLER-452 SYSTEM**

**SOLAR SYSTEM**

Mercury

Kepler-452b

Kepler-186f

Venus  Earth

Mars

# TRAPPIST-1, SYSTEM WITH 7 EARTH-LIKE EXOPLANETS

In addition to the three known planets discovered by the scientific community, four additional new planets were discovered orbiting the star system TRAPPIST-1. All planets in this multi-planetary system are surprisingly similar in size to Earth and at least three are within the "habitable zone" of its star system, an ultra-cool dwarf red star located 39 light years from Earth. In total, eight terrestrial telescopes as well as NASA's Hubble and Spitzer Space Telescopes where used in this search and discovery.

b          c          d          e          f          g          h

## CLOSE PROXIMITY

The seven planets of the TRAPPIST-1 system are very close to one another. Because the star is smaller and similar to Jupiter in size, the "habitable zone" is also close to the star.

## THE IDEAL CANDIDATE

If its existence can be confirmed, KOI-4878.01 will become the exoplanet most resembling Earth with an ESI of 98 percent. Detected by the Kepler Space Telescope, this extrasolar planet with a mass nearly identical to the Earth, orbits a sun-like yellow dwarf star every 449 days. In addition, it is located in the "habitable zone" of its system. Without knowing the composition of its atmosphere, it may meet all the requirements to support life even at 1,075 light years away.

## KEPLER-186F

**Artist's rendering of first exoplanet similar to Earth discovered in the "habitable zone" of a star system.**

## PROTECTED

Life on Earth would be impossible without the
presence of the atmosphere—the colorless,
odorless, invisible layer of gases that
surrounds us, giving us air to breathe and
protecting us from the Sun's harmful radiation.

## CHAPTER 3

# THE EARTH
## AND THE MOON

In the beginning, the Earth was an incandescent mass that slowly began to cool, allowing the continents to emerge and acquire their current form. Although many drastic changes took place during these early eras, our blue planet has still not stopped changing.

# THE BLUE PLANET

**The Earth is known as the blue planet because of the color of the oceans that cover two thirds of its surface.** This planet, the third planet from the Sun, is the only one where the right conditions exist to sustain life, something that makes the Earth special. It has liquid water in abundance, a mild temperature, and an atmosphere that protects it from objects that fall from outer space. The atmosphere also filters solar radiation thanks to its ozone layer. Slightly flattened at its poles and wider at its equator, the Earth takes 24 hours to revolve once on its axis.

## THE BIOSPHERE

Only a small part of the Earth is inhabited by living things: the surface, the oceans, the first 5 miles (8 km) of the atmosphere above the ground, and the area beneath the ground as far as plant roots reach. The biosphere makes up this tiny portion of the planet. Studying the biosphere helps to reveal the patterns by which different forms of life became established and the parameters that affect the distribution of species and ecosystems.

## 70%

OF THE EARTH'S SURFACE IS WATER. FROM SPACE, THE PLANET LOOKS BLUE.

## 23.5°

THIS IS THE INCLINATION OF THE EARTH'S AXIS FROM THE VERTICAL. AS THE EARTH ORBITS THE SUN, DIFFERENT REGIONS GRADUALLY RECEIVE MORE OR LESS SUNLIGHT, CAUSING THE FOUR SEASONS.

## THE EARTH'S MOVEMENTS

Night and day, summer and winter, new year and old year result from the Earth's various movements during its orbit of the Sun. The most important motions are the Earth's daily rotation from west to east on its own axis and its revolution around the Sun. (The Earth follows an elliptical orbit that has the Sun at one of the foci of the ellipse, so the distance to the Sun varies slightly over the course of a year.)

SUN

149,503,000 km
(93,500,000 mi)

**ROTATION:** The Earth revolves on its axis in 23 hours and 56 minutes.

**REVOLUTION:** It takes the Earth 365 days, 5 hours, and 57 minutes to travel once around the Sun.

**The Moon,** our only natural satellite, is four times smaller than the Earth and takes 27.32 days to orbit the Earth.

SOUTH POLE

**AXIS
INCLINATION**

NORTH
POLE

**ROTATION
AXIS**

## WATER STATES

### ① EVAPORATION

Because of the Sun's energy, the water evaporates, entering the atmosphere from oceans and, to a lesser extent, from lakes, rivers, and other sources on the continents.

### ② CONDENSATION

The Earth's winds transport moisture-laden air until weather conditions cause the water vapor to condense into clouds and eventually fall to the ground as rain or other forms of precipitation.

### ③ PRECIPITATION

The atmosphere loses water through condensation. Gravity causes rain, snow, and hail. Dew and frost directly alter the state of the surface they cover.

**CONVENTIONAL
PLANETARY
SYMBOL – EARTH**

### ESSENTIAL DATA

**Average distance from the Sun**
150 million km (93,5 million miles)

**Revolution around the Sun (Earth year)**
365.25 days

**Diameter at the equator**
12,756 km (7,930 mi)

**Orbital speed**
27.79 km/s (17 mi/s.)

**Mass***
1

**Gravity***
1

**Density**
3.2 ounces per cubic inch (5.52 g/cu cm)

**Average temperature**
15° C (59° F)

*In both cases, Earth = 1

**Moons**
1

**AXIAL TILT**

23.5°
One rotation
lasts 23.56 hours.

## GEOGRAPHIC COORDINATE

Thanks to the grid formed by the lines of latitude and longitude, the position of any object on the Earth's surface can be easily located by using the intersection of the Earth's equator and the Greenwich meridian (longitude 0°) as a reference point. This intersection marks the midpoint between the Earth's poles.

**0°
GREENWICH
MERIDIAN**

Northern
Hemisphere

Temperate
zone

**PARALLELS**
66.5° N Arctic Circle
23.5° N Tropic of Cancer

Tropical
zone

**0° EQUATOR**
23.5° S Tropic of Capricorn
66.5° S Antarctic Circle

Polar
zone

Southern
Hemisphere

# JOURNEY TO THE CENTER OF THE EARTH

**Earth is made up of different layers which in turn are made up of different elements** such as iron, nickel and rock in solid and in liquid form, in addition to fresh water, salt water and air. A cloud of gases encompasses our planet: the atmosphere. One of these gases – oxygen – allows the planet to sustain the majority of life.

## INNER LAYERS

We live on a rocky surface comprising, for the most part, oxygen and silica. Underneath, there is the mantle; the rocks here are much heavier. Beneath the mantle are the inner and outer crusts. The former comprises constantly boiling liquid metal. The latter, which is solid due to the effects of pressure, is the densest part of the planet.

### HOW FAR HUMANKIND HAS REACHED

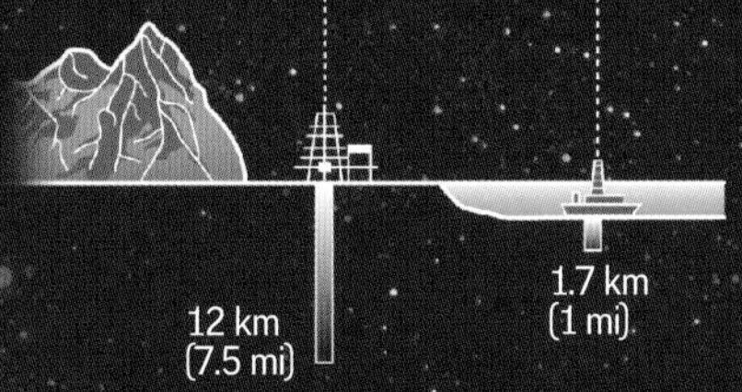

Mount Everest 8,850 m (29,029 ft)

Offshore drilling

Ocean drilling

12 km (7.5 mi)

1.7 km (1 mi)

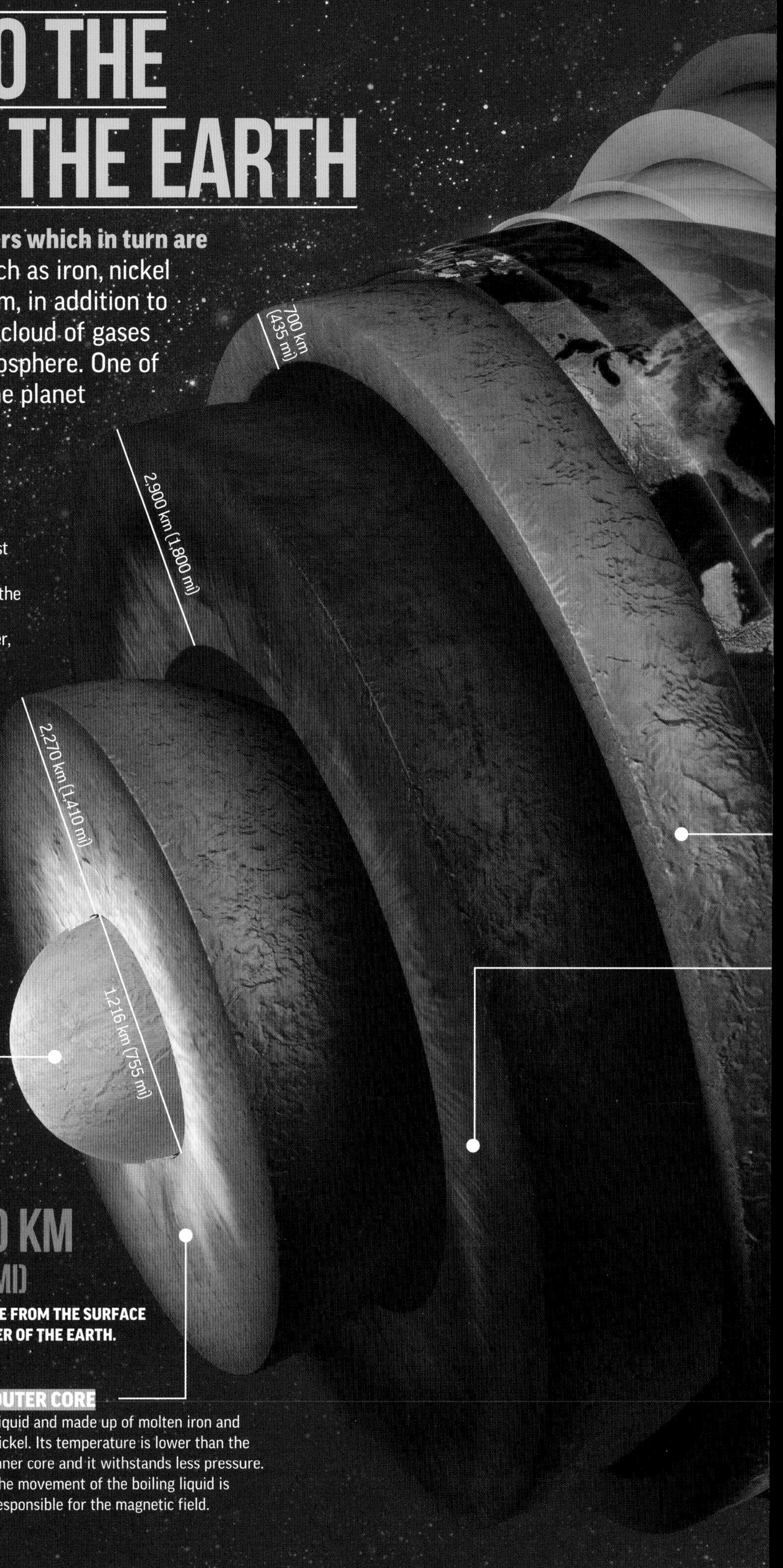

**INNER CORE**

Comprising the same metals as the outer core but, unlike the layer that covers it and despite the extreme temperature, the center is solid due to the enormous pressures exerted on it, causing it to compress as a result.

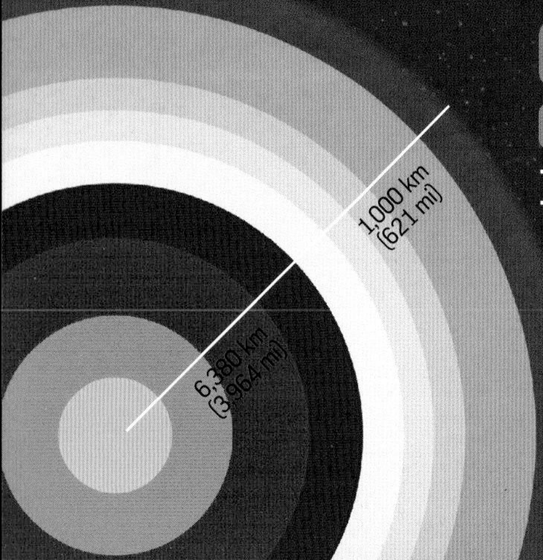

## 6,380 KM
### (3,964 MI)

**THE DISTANCE FROM THE SURFACE TO THE CENTER OF THE EARTH.**

**OUTER CORE**

Liquid and made up of molten iron and nickel. Its temperature is lower than the inner core and it withstands less pressure. The movement of the boiling liquid is responsible for the magnetic field.

700 km (435 mi)

2,900 km (1,800 mi)

2,270 km (1,410 mi)

1,216 km (755 mi)

1,000 km (621 mi)

6,380 km (3,964 mi)

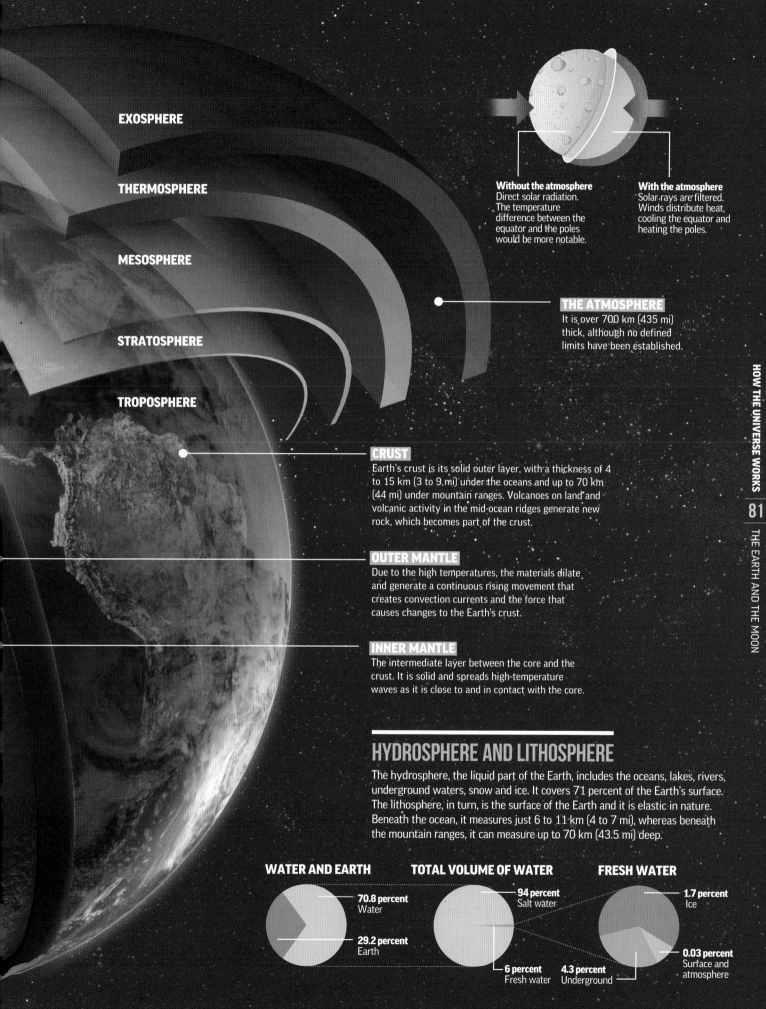

EXOSPHERE

THERMOSPHERE

MESOSPHERE

STRATOSPHERE

TROPOSPHERE

**Without the atmosphere**
Direct solar radiation.
The temperature
difference between the
equator and the poles
would be more notable.

**With the atmosphere**
Solar rays are filtered.
Winds distribute heat,
cooling the equator and
heating the poles.

**THE ATMOSPHERE**
It is over 700 km (435 mi)
thick, although no defined
limits have been established.

**CRUST**
Earth's crust is its solid outer layer, with a thickness of 4
to 15 km (3 to 9 mi) under the oceans and up to 70 km
(44 mi) under mountain ranges. Volcanoes on land and
volcanic activity in the mid-ocean ridges generate new
rock, which becomes part of the crust.

**OUTER MANTLE**
Due to the high temperatures, the materials dilate
and generate a continuous rising movement that
creates convection currents and the force that
causes changes to the Earth's crust.

**INNER MANTLE**
The intermediate layer between the core and the
crust. It is solid and spreads high-temperature
waves as it is close to and in contact with the core.

# HYDROSPHERE AND LITHOSPHERE

The hydrosphere, the liquid part of the Earth, includes the oceans, lakes, rivers,
underground waters, snow and ice. It covers 71 percent of the Earth's surface.
The lithosphere, in turn, is the surface of the Earth and it is elastic in nature.
Beneath the ocean, it measures just 6 to 11 km (4 to 7 mi), whereas beneath
the mountain ranges, it can measure up to 70 km (43.5 mi) deep.

**WATER AND EARTH**

**70.8 percent**
Water

**29.2 percent**
Earth

**TOTAL VOLUME OF WATER**

**94 percent**
Salt water

**6 percent**
Fresh water

**4.3 percent**
Underground

**FRESH WATER**

**1.7 percent**
Ice

**0.03 percent**
Surface and
atmosphere

# THE EARTH, A HUGE MAGNET

**The Earth behaves like a giant bar magnet and has a magnetic field with two poles.** It is likely that the Earth's magnetism results from the motion of the iron and nickel in its electroconductive core. Another probable origin of the Earth's magnetism lies in the convection currents caused by the heat of the core. The Earth's magnetic field has varied over the course of time. During the last five million years, more than 20 reversals have taken place. The most recent one occurred 700,000 years ago.

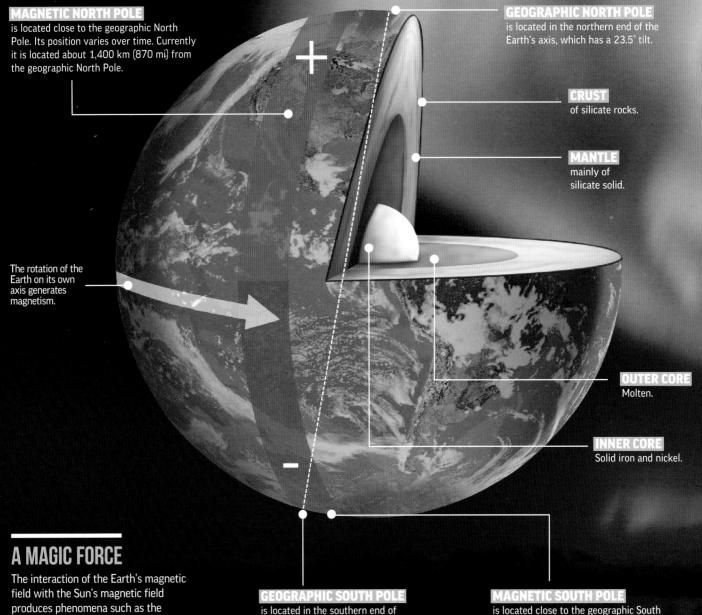

**MAGNETIC NORTH POLE**
is located close to the geographic North Pole. Its position varies over time. Currently it is located about 1,400 km (870 mi) from the geographic North Pole.

**GEOGRAPHIC NORTH POLE**
is located in the northern end of the Earth's axis, which has a 23.5° tilt.

**CRUST**
of silicate rocks.

**MANTLE**
mainly of silicate solid.

The rotation of the Earth on its own axis generates magnetism.

**OUTER CORE**
Molten.

**INNER CORE**
Solid iron and nickel.

**GEOGRAPHIC SOUTH POLE**
is located in the southern end of the Earth's axis.

**MAGNETIC SOUTH POLE**
is located close to the geographic South Pole. Its position varies over time. Currently it is located about 2,750 km (1,700 mi) from the geographic South Pole.

## A MAGIC FORCE

The interaction of the Earth's magnetic field with the Sun's magnetic field produces phenomena such as the aurora borealis and australis; the interaction can also cause interference in radio-wave transmissions.

# MAGNETOSPHERE

The invisible lines of force that form around the Earth. It has an ovoid shape and extends 60,000 km (37,000 mi) from the Earth. Among other things, it protects the Earth from harmful particles radiated by the Sun.

The atmosphere reaches 900 km (560 mi).

Solar wind with charged atomic particles

The deformation of the magnetosphere is caused by the action of electrically charged particles streaming from the Sun.

The Van Allen belts are bands of ionized atomic particles.

## PLANETARY AND SOLAR MAGNETISM

The planets in the Solar System have various magnetic fields with varying characteristics. The four giant planets possess stronger magnetic fields than the Earth.

**NEPTUNE**   **URANUS**   **SATURN**   **JUPITER**   **MARS**   **EARTH**   **VENUS**   **MERCURY**   **SUN**

It is believed that in the past its magnetic field was stronger.

It is the only planet in the Solar System that does not have a magnetic field.

The gases that flow from the Sun's corona produce a magnetic field around it.

# AURORA BOREALIS

The interaction between the magnetic fields of the Sun and the Earth and the charged solar particles that arrive at the Earth generate the beautiful boreal and austral auroras close to the poles.

1. The Sun's magnetic field expels particles into space: the solar wind.

2. The solar wind is diverted by the Earth's magnetic field.

3. Some particles (protons and neutrons) are guided by the Earth's magnetic field toward the poles.

4. The particles crash with oxygen and nitrogen atoms present in the atmosphere, and the atoms are elevated to an excited state. As a result of their excited state, the atoms emit energy in the form of light.

Solar wind

Magnetic fields

# SURFACE AND MOVEMENTS OF THE MOON

**It is believed that the Moon was created when a Mars-sized body crashed into Earth while it was still in formation.** The expelled material scattered around Earth and, over time, it joined together to form the Moon. It is our planet's only natural satellite and its gravitational pull influences the tides. Depending on its position, its gravitational pull over the Earth's bodies of water is greater or smaller.

## LUNAR MOVEMENTS

For each terrestrial orbit, the Moon spins on its own axis. As a result, the same side always faces Earth.

**Lunar month**
It takes 29.53 days to complete its phase.

**Sidereal month**
It takes 27.32 days to orbit Earth.

**Hidden face**
It was not until 1953, when the Luna 3 probe photographed it, that the Moon's hidden face was seen for the first time.

Visible face

Moon

Earth

Lunar orbit

**CONVENTIONAL PLANETARY SYMBOL – MOON**

## ESSENTIAL DATA

**Distance from Earth**
384,400 km (238,855 mi)

**Orbit around Earth**
27.3 days

**Diameter at the equator**
3,476 km (2,160 mi)

**Orbital speed**
1.02 km/sec (0.63 mi/sec)

**Mass (Earth = 1)**
0.01

**Gravity (Earth = 1)**
0.17

**Density**
3.34 g/cm³

**Average temperature**
150° C (302° F) – day
-100° C (-148° F) – night

**Solar orbit (Earth year)**
365.25 days

**AXIAL TILT**

5.14°
One rotation
takes 27.32 days.

**ARISTARCHUS**
Is the brightest point on the Moon.

**OCEANUS PROCELLARUM**
Is the biggest sea.

**COPERNICUS**
Crater measuring 93 km (58 mi) in diameter.

**MONTES APENNINUS**
One of the most important mountain ranges on the Moon.

Schickard

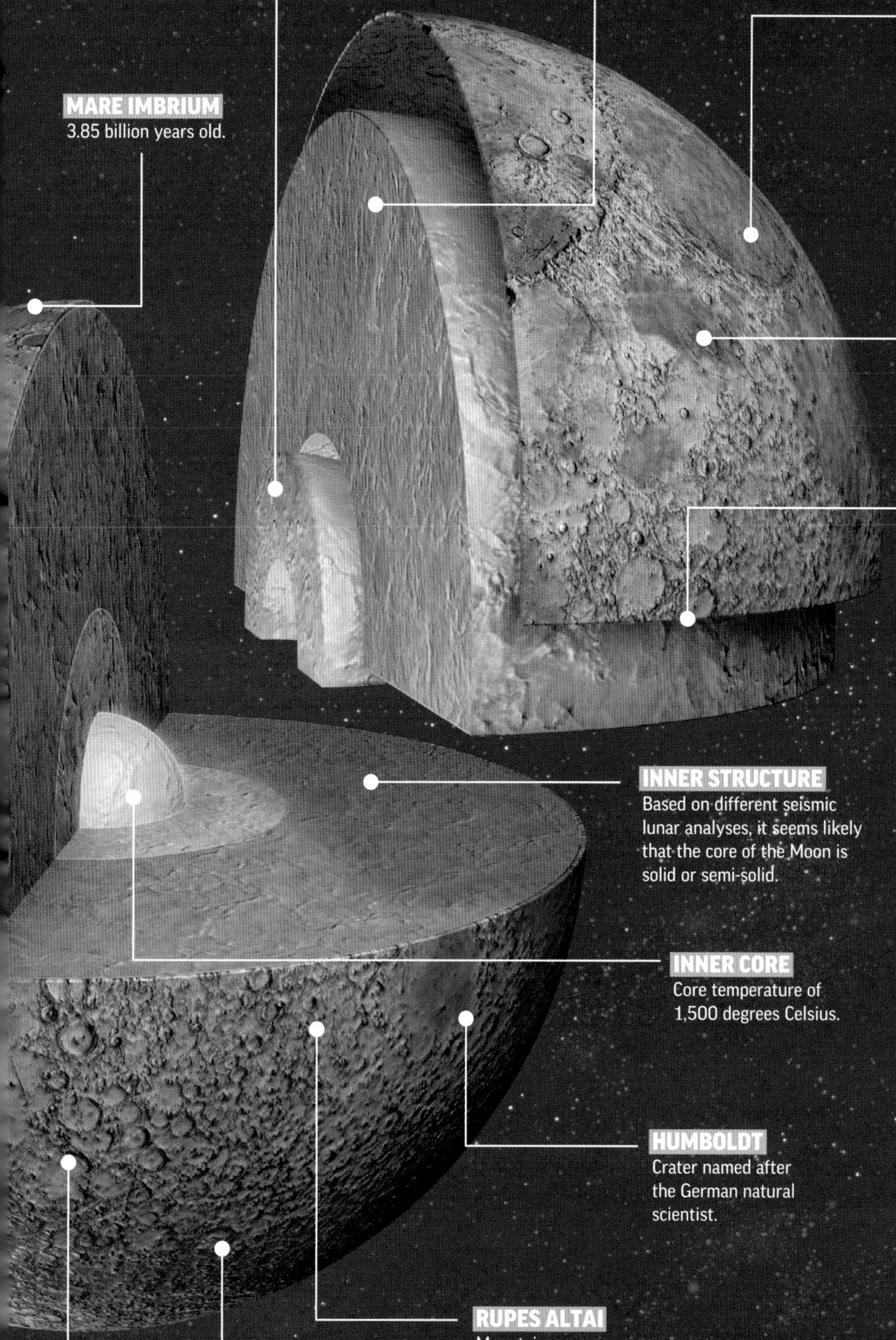

**OUTER CORE**
Partially molten.

**ROCKY CRUST**
Less than half the depth of Earth.

**MARE CRISIUM**
Measuring 450 km by 563 km, (280 x 350 mi). It is scarred by large craters.

**MARE IMBRIUM**
3.85 billion years old.

**Mare Tranquillitatis**

**CRUST**
Rocky, granite-like surface with 20 m (66 ft) of lunar dust known as 'regolith.'

**INNER STRUCTURE**
Based on different seismic lunar analyses, it seems likely that the core of the Moon is solid or semi-solid.

# THE SURFACE OF THE MOON

Ancient astronomers deduced that the dark patches on the Moon that can be seen with the naked eye were seas. These dark areas contrast with the light areas (highlands with a higher number of craters).

## MOUNTAIN RANGES
Formed from material expelled from the crater after a meteorite impacted against the surface of the Moon.

**INNER CORE**
Core temperature of 1,500 degrees Celsius.

## CRATERS
Different craters measure between 1 m and 1,000 km (3 ft and 621 mi) in diameter. They were formed as a result of collisions with meteorites.

**HUMBOLDT**
Crater named after the German natural scientist.

## SEAS
They cover 16 percent of the Moon's surface and were formed by lava channels. Today, there is no volcanic activity on the Moon. However, this was not always the case.

**RUPES ALTAI**
Mountain range measuring 1,800 m (5,905 ft) in height.

**Tycho**
100 million years old.

**Maginius**

# THE MOON AND THE TIDES

**The water is held on Earth by gravity, but is also attracted by the gravity of the Moon and the Sun.** This effect combined with the rotation of the Earth, which lasts 24 hours, causes the tides to rise and fall. That is, the oceans move from side to side of the Earth as if it were a giant bowl that we are shaking. The peak that reaches the tide is called high tide and, the lowest, low tide.

## PHASES OF THE MOON

The difference between high tide and low tide varies with the phases of the Moon. Water that is closer to the Moon feels the force of gravity with greater intensity than water on the opposite side of the Earth and thus the movements of the tides follow the Moon as it orbits the Earth.

### REFERENCES

Gravitational pull of the Moon.

Gravitational pull of the Sun.

Influence in the tide caused by the gravitational force of the Sun.

Influence in the tide caused by the gravitational pull of the Moon.

## PERIODICITY

In the majority of coasts one observes two high tides and two low tides per day. The average interval between two high tides and two low tides is 12 hours and 25 minutes. This corresponds to half the time it takes the Earth to rotate with respect to the line which unites the Earth and the Moon.

**Lunar orbit**

**Earth orbit**

**Sagittarius Dwarf**

Moon

(1) **NEW MOON**

Spring-tide

**As the Sun and the Moon are aligned, they result in the highest high tides and the lowest low tides.**

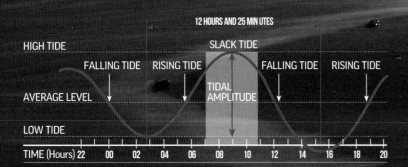

12 HOURS AND 25 MINUTES

| HIGH TIDE | | | SLACK TIDE | | |
| FALLING TIDE | RISING TIDE | | | FALLING TIDE | RISING TIDE |

TIDAL AMPLITUDE

AVERAGE LEVEL

LOW TIDE

TIME (Hours) 22  00  02  04  06  08  10  12  14  16  18  20

IN THE WORLD

Daily    Mixed    Half daily

## ② FIRST QUARTER

In a right angle to the Earth, the Moon and Sun generate the lowest high tides and highest low tides.

The Sun and Moon align once again, and the Sun counteracts the pull of the Moon: second spring tide.

## RANGE OF TIDE

The height difference between tides is usually about 80 cm (2.62 ft), but it varies greatly by region. In some parts of the English Channel it surpasses 10 m (32.81 ft) while in the Baltic Sea, it is imperceptible.

## ④ LAST QUARTER

**Neap tide**
The Sun and Moon form a right angle again, causing the second neap tide.

## 2 DAYS

SPRING AND NEAP TIDES OCCUR A FEW DAYS AFTER THE PHASES OF THE MOON, BECAUSE OF THE FRICTION AND INERTIA OF THE WATER.

### THE SUN

The Sun's gravity also influences the movements of the tides, although in its case it represents only 46.6% of the influence of the Moon.

## TIDAL ENERGY

The force generated by tides is used to produce electricity. Some of the world's largest plants are La Rance (France), Sihwa Lake (South Korea), Tidal Lagoon (UK) and Annapolis Royal (Canada).

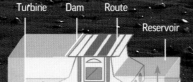

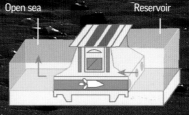

Turbine   Dam   Route   Reservoir   Open sea   Reservoir

Bed

### HIGH TIDE

The water enters the reservoir at 18,000 m³/s and makes the turbines that generate electricity work.

### LOW TIDE

At low tide, the water flows to the estuary and again drives the turbines that generate electricity.

# ECLIPSES

At least four times a year, the centers of the Moon, the Sun and the Earth fully align, resulting in one of the most attractive astronomical events to the casual observer: eclipses. Solar eclipses also provide astronomers with an amazing opportunity for scientific investigation.

## SOLAR ECLIPSE

Solar eclipses occur when the Moon passes directly between the Sun and the Earth, casting a shadow along a path on the Earth's surface. The central cone of the shadow is called the umbra, and the area of partial shadow around it is called the penumbra. Viewers in the regions where the umbra falls on the Earth's surface see the Moon's disk completely obscure the Sun—a total solar eclipse. Those watching from the surrounding areas that are located in the penumbra see the Moon's disk cover only part of the Sun—a partial solar eclipse.

**TOTAL LUNAR ECLIPSE, SEEN FROM THE EARTH**
The orange color comes from sunlight that has been refracted and colored by the Earth's atmosphere.

**ANNULAR SOLAR ECLIPSE SEEN FROM EARTH**

### DISTANCE FROM THE SUN TO THE EARTH

# 400
**TIMES GREATER THAN THE DISTANCE FROM THE EARTH TO THE MOON.**

### ALIGNMENT

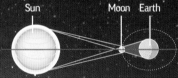

Sun          Moon  Earth

During a solar eclipse, astronomers take advantage of the blocked view of the Sun in order to use devices designed to study the Sun's atmosphere.

### TYPES OF ECLIPSES

**TOTAL**
The Moon is positioned between the Sun and Earth, within the shadow zone.

**ANNULAR**
The diameter of the Moon is smaller than that of the Sun, and part of the Sun can be seen.

**PARTIAL**
The Moon does not fully cover the Sun, which appears as a crescent.

### SUN'S APPARENT SIZE

# 400
**TIMES LARGER THAN THE MOON**

SUNLIGHT

# LUNAR ECLIPSE

When Earth passes between the Moon and the Sun, the resulting phenomenon is a lunar eclipse, which may be total, partial or penumbral. A totally eclipsed Moon takes on a characteristic reddish color, as the light is refracted by the Earth's atmosphere. When part of the Moon is within the shadow zone, and the rest within the penumbral zone, the result is a partial eclipse.

## ALIGNMENT

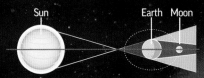

Sun    Earth Moon

During an eclipse, the Moon is not completely black; it assumes an ochre color.

## TYPES OF ECLIPSES

**TOTAL**
The Moon is completely within the shadow zone.

**PARTIAL**
The Moon is only partially within the shadow zone.

**PENUMBRAL**
The Moon is within the penumbral zone.

Lunar orbit

Shadow zone

FULL MOON
TOTAL ECLIPSE

PARTIAL ECLIPSE

PENUMBRAL ECLIPSE

Penumbral zone

NEW MOON
TOTAL ECLIPSE

Shadow zone

Penumbral zone

EARTH

Earth's orbit

## THE ECLIPSE CYCLE

Eclipses are repeated every 223 lunar months, or every 18 years and 11 days. These periods are called 'saros.'

**ECLIPSES IN A YEAR**

| 2 | 7 | 4 |
|---|---|---|
| Minimum | Maximum | Average |

**ECLIPSES IN A SAROS**

| 41 | 29 | 70 |
|---|---|---|
| of the Sun | of the Moon | Total |

## OBSERVATION FROM EARTH

A black, polymer filter, with an optical density of 5.0, produces a clear orange image of the Sun. Prevents retinal burns.

## SOLAR ECLIPSES

are different for each local observer.

**MAXIMUM DURATION**

**7.5** MINUTES

## LUNAR ECLIPSES

are the same for all observers.

**MAXIMUM DURATION**

**107** MINUTES

## LAST ECLIPSES

| OF THE SUN | | | | | | | | | | | | | | | | | | |
|---|---|---|---|---|---|---|---|---|---|---|---|---|---|---|---|---|---|---|
| 3/29 Total | 9/22 Total | 3/19 Partial | 9/11 Partial | 2/07 Total | 1/26 Total | 7/22 Total | 1/15 Total | 7/11 Total | 1/4 Partial | 11/25 Partial | 5/20 Annular | 11/13 Annular | 5/10 Annular | 11/3 Hybrid | 4/29 Annular | 10/23 Partial | 3/20 Total | 9/13 Partial |

| 2006 | 2007 | 2008 | 2009 | 2010 | 2011 | 2012 | 2013 | 2014 | 2015 | 2016 |

| OF THE MOON | | | | | | | | | | | | | | | | | | | |
|---|---|---|---|---|---|---|---|---|---|---|---|---|---|---|---|---|---|---|---|
| 3/14 Partial | 9/07 Partial | 3/03 Total | 8/28 Total | 2/21 Total | 8/16 Partial | 2/9 Partial | 7/7 Partial | 6/26 Partial | 12/21 Total | 6/15 Total | 12/10 Total | 6/4 Partial | 12/28 Partial | 4/25 Partial | 10/18 Partial | 4/15 Total | 10/08 Total | 4/4 Total | 9/28 Total |

# TRAVERSING TIME

Geologists and paleontologists use many sources to reconstruct the Earth's history. The analysis of rocks, minerals, and fossils found on the Earth's surface provides data about the deepest layers of the planet's crust and reveals both climatic and atmospheric changes that are often associated with catastrophes. Craters caused by the impact of meteorites and other bodies on the surface of the Earth also reveal valuable information about the history of the planet.

(2) **Collision and fusion**
The heavy elements migrate.

## COMPLEX STRUCTURE

The formation of the interior cosmic materials began to accumulate, forming a growing celestial body, the precursor of the Earth. High temperatures combined with gravity caused the heaviest elements to migrate to the center of the planet and the lighter ones to move toward the surface. Under a rain of meteors, the external layers began to consolidate and form the Earth's crust. In the center, metals such as iron concentrated into a red-hot nucleus.

(1) **Small bodies and dust** accumulate to become the size of an asteroid.

The oldest minerals, such as zircon, form.

The oldest rocks metamorphose, forming gneiss.

**1,100**
**RODINIA, AN EARLY SUPERCONTINENT, FORMS.**

A meteorite falls in Sudbury, Ontario, Canada.

Age in millions of years **4,600**

**2,500**

| ERA | HADEAN | PROTEROZOIC |
| PERIOD | PREGEOLOGIC | PRECAMBRIAN |
| EPOCH | | |

**CLIMATE**

Consolidation begins under a rain of meteors.

The Earth cools and the first ocean is formed.

**2,500**
**THE EARTH UNDERGOES THE FIRST OF ITS MASSIVE GLOBAL COOLING EVENTS (GLACIATIONS).**

### ELEMENTS PRESENT ACCORDING TO THE TABLE
Existing in different combinations, the crust of the Earth contains the same elements today as those that were present when the planet was formed. The most abundant element in the crust is oxygen, which bonds with metals and nonmetals to form different compounds.

**800** SECOND GLACIATION
**600** LAST MASSIVE GLACIATION

**LIFE**

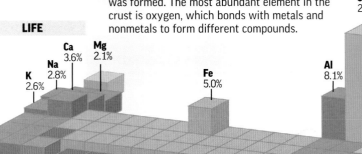

K 2.6%
Na 2.8%
Ca 3.6%
Mg 2.1%
Fe 5.0%
Al 8.1%
Si 27.7%
O 46.6%

- Metals
- Transition metals
- Nonmetals
- Noble gases
- Lanthanide series
- Actinide series

### THE FIRST ANIMALS
Among the most mysterious fossils of the Precambrian Period are the remains of the Ediacaran fauna, the Earth's first-known animals. They lived at the bottom of the ocean. Many were round and reminiscent of jellyfish, while others were flat and sheetlike.

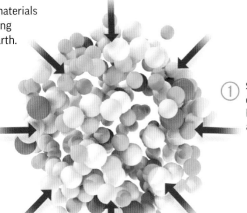

## Metallic core

(3) The light elements form the mantle.

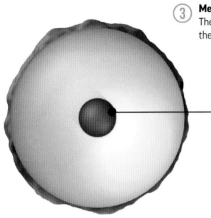

**THE CORE**
The Earth's core is extremely hot and is made mostly of iron and nickel.

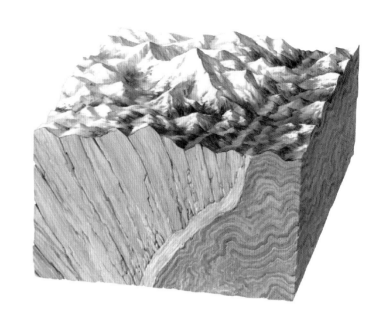

# MOUNTAINS

**ARE EXTERNAL FOLDS OF THE CRUST PRODUCED BY EXTREMELY POWERFUL FORCES OCCURRING INSIDE THE EARTH.**

## 542

**THE SUPERCONTINENT PANOTIA FORMS, CONTAINING PORTIONS OF PRESENT-DAY CONTINENTS. NORTH AMERICA SEPARATES FROM PANOTIA.**

## OROGENIES

Geological history recognizes long periods (lasting millions of years) of intense mountain formation called orogenies. Each orogeny is characterized by its own particular materials and location.

The first major orogeny (Caledonian folding) begins. Gondwana moves toward the South Pole.

Laurentia and Baltica converge, creating the Caledonian range. Gneiss forms on the coast of Scotland.

The region that will become North America moves toward the Equator, thus initiating the development of the most important carboniferous formations. Gondwana moves slowly; the ocean floor spreads at a similar speed.

The fragments of continents combine to form a single continent called Pangea.

The Appalachian Mountains form. The formation of slate through sedimentation is at its peak.

Baltica and Siberia clash, forming the Ural Mountains.

Eruptions of basalt occur in Siberia.

| 542 | 488.3 | 443.7 | 416 | 359.2 | 299 |

▶ **PALEOZOIC** THE ERA OF PRIMITIVE LIFE

▶ **CAMBRIAN**   ▶ **ORDOVICIAN**   ▶ **SILURIAN**   ▶ **DEVONIAN**   ▶ **CARBONIFEROUS**   ▶ **PERMIAN**

Temperatures fall. The level of carbon dioxide ($CO_2$) in the atmosphere is 16 times higher than it is today.

It is thought that the Earth's atmosphere contained far less carbon dioxide during the Ordovician than today. Temperatures fluctuate within a range similar to what we experience today.

By this period, vertebrates with mandibles, such as the placoderms, osteichthyans (bony fish), and acanthodians, have already emerged.

Temperatures were typically warmer than today, and oxygen ($O_2$) levels attained their maximum.

Hot, humid climates produce exuberant forests in swamplands.

The largest carbon deposits we observe today form where forests previously existed.

**TRILOBITES**
Marine arthropods with mineralized exoskeletons

## THE CAMBRIAN EXPLOSION

Fossils from this time attest to the great diversity of marine animals and the emergence of different types of skeletal structures, such as those found in sponges and trilobites.

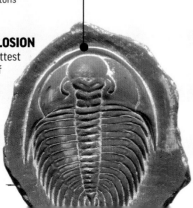

**SILURIAN**
One of the first pisciform vertebrates, an armored fish without mandibles

The rocks of this period contain an abundance of fish fossils.

Areas of solid ground are populated by gigantic ferns.

Amphibians diversify and reptiles originate from one amphibian group to become the first amniotes. Winged insects such as dragonflies emerge.

Palm trees and conifers replace the vegetation from the Carboniferous Period.

## MASS EXTINCTION

Near the end of the Permian Period, an estimated 95% of marine organisms and over two thirds of terrestrial ones perish in the greatest known mass extinction.

# IMPACT FROM THE OUTSIDE

It is believed that a large meteor fell on Chicxulub, on the Yucatán Peninsula (Mexico), about 65 million years ago. The impact caused an explosion that created a cloud of ash mixed with carbon rocks. When the debris fell back to Earth, some experts believe it caused a great global fire.

The heat caused by the expansion of fragments from the impact together with the greenhouse effect, brought about by the spreading of ashes in the stratosphere, provoked a series of climatic changes. It is believed that this process resulted in the extinction of the dinosaurs.

## 100 KM
### (62 MI)

**THE DIAMETER OF THE CRATER PRODUCED BY THE IMPACT OF THE METEOR ON THE YUCATÁN PENINSULA. IT IS NOW BURIED UNDER ALMOST 3,2 KM (2 MI) OF LIMESTONE.**

Gondwana reappears.

Africa separates from South America, and the South Atlantic Ocean appears.

**FORMATION OF MOUNTAIN CHAINS**

**CENTRAL ROCKY MOUNTAINS**

**ALPS**

**HIMALAYAS**

| 251 | 199.6 | 145.5 | 65.5 |
|---|---|---|---|

| MESOZOIC THE ERA OF REPTILES | | | CENOZOIC THE AGE OF MAMMALS |
|---|---|---|---|
| TRIASSIC | JURASSIC | CRETACEOUS | PALEOGENE |
| | | | PALEOCENE / EOCENE |

Carbon dioxide levels increase. Average temperatures are higher than today.

The level of oxygen (O$_2$) in the atmosphere is much lower than today.

**THE AGE OF FLOWERING PLANTS**
At the end of the Cretaceous Period, the first angiosperms—plants with protected seeds, flowers, and fruits—appear.

The global average temperature is at least 17° C (62° F). The ice layer covering Antarctica later thickens.

Proliferation of insects

Appearance of dinosaurs

The first mammals evolve from a group of reptiles called Therapsida.

Birds emerge.

The dinosaurs undergo adaptive radiation.

**ALLOSAURUS**
This carnivore measured 12 m (39 ft) long.

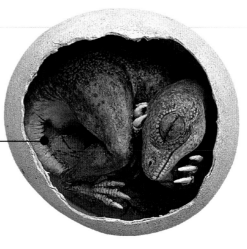

**ANOTHER MASS EXTINCTION**
Toward the end of the Cretaceous Period, about 50% of existing species disappear. The dinosaurs, the large marine reptiles (such as the Plesiosaurs), the flying creatures of that period (such as the Pterosaurs), and the ammonites (cephalopod mollusks) disappear from the Earth. At the beginning of the Cenozoic Era, most of the habitats of these extinct species begin to be occupied by mammals.

Panthalassa

Pangea

**(1) 290 million years ago.** The supercontinent called Pangea formed. An immense ocean called Panthalassa surrounded it.

Laurasia

Tethys sea

Gondwana

**(2) 250 million years ago.** The Tethys Sea slowly split Pangea, creating two continents, known as Laurasia and Gondwana.

North and South America are joined at the end of this time period. The formation of Patagonia concludes, and an important overthrust raises the Andes mountain range.

The African Rift Zone and the Red Sea open up. The Indian protocontinent collides with Eurasia.

**(3) 163 million years ago.** Gondwana split, forming Africa and South America as the southern Atlantic Ocean was created.

**(4) 60 million years ago.** The northern Atlantic Ocean slowly separated, completing the formation of Europe and North Africa.

23

250 million years ago, India, Africa, Australia, and Antarctica were part of the same continent. When tectonic plates rub against each other, land and oceanic crust earthquakes occur. Where the plates separate, a rift forms. The mid-ocean ridges that run beneath the oceans are formed by lava that emerges from the rifts between tectonic plates. Where plates collide, a process called subduction takes place, in which the rocks of the oceanic floor are drawn under the continent and melt, reemerging in the form of volcanoes.

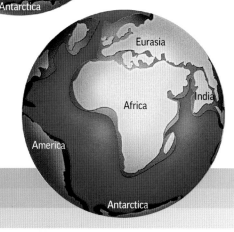

Laurasia

Africa

India

South America

Antarctica

### TECTONIC PLATES
The surface of the Earth is shaped by tectonic plates. There are eight major plates, some of which even encompass entire continents. The plates' borders are marked by ocean trenches, cliffs, chains of volcanoes, and earthquake zones.

Eurasia

Africa

India

America

Antarctica

---

Temperatures drop to levels similar to those of today. The lower temperatures cause forests to shrink and grasslands to expand.

### THE LAST GLACIATION
The most recent period of glaciation begins three million years ago and intensifies at the beginning of the Quaternary Period. North Pole glaciers advance, and much of the Northern Hemisphere becomes covered in ice.

### HUMAN BEINGS APPEAR ON EARTH
Although the oldest hominid fossils (*Sahelanthropus*) date back to seven million years ago, it is believed that modern humans emerged in Africa at the end of the Pleistocene. Humans migrated to Europe 100,000 years ago, although settling there was difficult because of the glacial climate. According to one hypothesis, our ancestors reached the American continent about 10,000 years ago by traveling across the area now known as the Bering Strait.

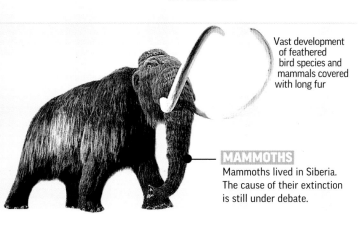

Vast development of feathered bird species and mammals covered with long fur

**MAMMOTHS**
Mammoths lived in Siberia. The cause of their extinction is still under debate.

# UNDER CONSTRUCTION

**Our planet is not a dead body, complete and unchanging. It is an ever-changing system whose activity we experience all the time: volcanoes erupt, earthquakes occur, and new rocks emerge on the Earth's surface.** All these phenomena, which originate in the interior of the planet, are studied in a branch of geology called internal geodynamics. This science analyzes processes, such as continental drift and isostatic movement, which originate with the movement of the crust and result in the raising and sinking of large areas. The movement of the Earth's crust also generates the conditions that form new rocks. This movement affects magmatism (the melting of materials that solidify to become igneous rocks) and metamorphism (the series of transformations occurring in solid materials that give rise to metamorphic rocks).

## MAGMATISM

Magma is produced when the temperature in the mantle or crust reaches a level at which minerals with the lowest fusion point begin to melt. Because magma is less dense than the solid material surrounding it, it rises, and in so doing it cools and begins to crystallize. When this process occurs in the interior of the crust, plutonic or intrusive rocks, such as granite, are produced. If this process takes place on the outside, volcanic or effusive rocks, such as basalt, are formed.

## METAMORPHISM

An increase in pressure and/or temperature causes rocks to become plastic and their minerals to become unstable. These rocks then chemically react with the substances surrounding them, creating different chemical combinations and thus causing new rocks to form. These rocks are called metamorphic rocks. Examples of this type of rock are marble, quartzite, and gneiss.

**OUTER CRUST**
Volcanic rocks.

**INNER CRUST**
Plutonic rocks.

Crust

Magmatic Chamber

Oceanic Plate

Sea Level

100 km (62 mi)

200 km (124 mi)

Convective Currents

Asthenosphere

**PRESSURE**
This force gives rise to new metamorphic rocks, as older rocks fuse with the minerals that surround them.

**TEMPERATURE**
High temperatures make the rocks plastic and their minerals unstable.

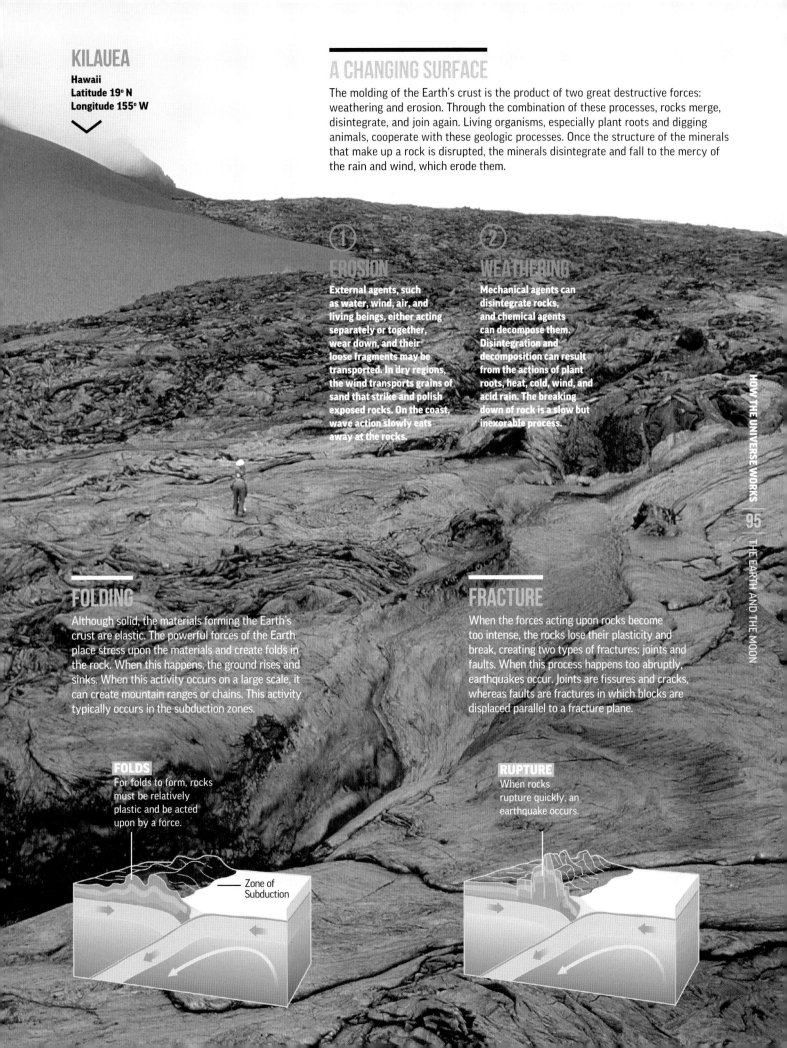

# A CHANGING SURFACE

The molding of the Earth's crust is the product of two great destructive forces: weathering and erosion. Through the combination of these processes, rocks merge, disintegrate, and join again. Living organisms, especially plant roots and digging animals, cooperate with these geologic processes. Once the structure of the minerals that make up a rock is disrupted, the minerals disintegrate and fall to the mercy of the rain and wind, which erode them.

## ① EROSION

External agents, such as water, wind, air, and living beings, either acting separately or together, wear down, and their loose fragments may be transported. In dry regions, the wind transports grains of sand that strike and polish exposed rocks. On the coast, wave action slowly eats away at the rocks.

## ② WEATHERING

Mechanical agents can disintegrate rocks, and chemical agents can decompose them. Disintegration and decomposition can result from the actions of plant roots, heat, cold, wind, and acid rain. The breaking down of rock is a slow but inexorable process.

## FOLDING

Although solid, the materials forming the Earth's crust are elastic. The powerful forces of the Earth place stress upon the materials and create folds in the rock. When this happens, the ground rises and sinks. When this activity occurs on a large scale, it can create mountain ranges or chains. This activity typically occurs in the subduction zones.

### FOLDS

For folds to form, rocks must be relatively plastic and be acted upon by a force.

Zone of Subduction

## FRACTURE

When the forces acting upon rocks become too intense, the rocks lose their plasticity and break, creating two types of fractures: joints and faults. When this process happens too abruptly, earthquakes occur. Joints are fissures and cracks, whereas faults are fractures in which blocks are displaced parallel to a fracture plane.

### RUPTURE

When rocks rupture quickly, an earthquake occurs.

# SCORCHING FLOW

**Most of the Earth's interior is in a liquid and incandescent state at extremely high temperatures.** This vast mass of molten rock contains dissolved crystals and water vapor, among other gases, and it is known as magma. When part of the magma rises toward the Earth's surface, mainly through volcanic activity, it is called lava. As soon as it reaches the surface of the Earth or the ocean floor, the lava starts to cool and solidify into different types of rock, according to its original chemical composition. This is the basic process that formed the surface of our planet, and it is the reason the Earth's surface is in constant flux. Scientists study lava to understand our planet better.

## STREAMS OF FIRE

Lava is at the heart of every volcanic eruption. The characteristics of lava vary, depending on the gases it contains and its chemical composition. Lava from an eruption is loaded with water vapor and gases such as carbon dioxide, hydrogen, carbon monoxide and sulfur dioxide. As these gases are expelled, they burst into the atmosphere, where they create a turbulent cloud that sometimes discharges heavy rains. Fragments of lava expelled and scattered by the volcano are classified as bombs, cinders and ash. Some large fragments fall back into the crater. The speed at which lava travels depends to a great extent on the steepness of the sides of the volcano. Some lava flows can reach 145 km (90 mi) in length and attain speeds of up to 50 km/h (30 mph).

### INTENSE HEAT

Lava can reach temperatures above 1,200° C (2,200° F). The hotter the lava, the more fluid it is. When lava is released in great quantities, it forms rivers of fire. The lava's advance is slowed down as the lava cools and hardens.

## MINERAL COMPOSITION

Lava contains a high level of silicates, light rocky minerals that make up 95 percent of the Earth's crust. The second most abundant substance in lava is water vapor. Silicates determine lava's viscosity, that is, its capacity to flow. Variations in viscosity have resulted in one of the most commonly used classification systems of lava: basaltic, andesitic, and rhyolitic, in order from least to greatest silicate content. Basaltic lava forms long rivers, such as those that occur in typical Hawaiian volcanic eruptions, whereas rhyolitic lava tends to erupt explosively because of its poor fluidity. Andesitic lava, named after the Andes mountains, where it is commonly found, is an intermediate type of lava of medium viscosity.

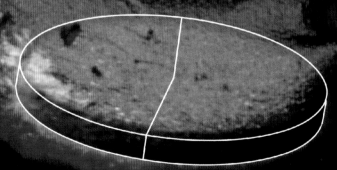

**48** % OTHER CONTENT

**52** % SILICATES

# ROCK CYCLE

Once it cools, lava forms igneous rock. This rock, subjected to weathering and natural processes such as metamorphism and sedimentation, will form other types of rocks that, when they sink back into the Earth's interior, again become molten rock. This process takes millions of years and is known as the rock cycle.

## SEDIMENTARY ROCK

Rock formed by eroded and compacted materials.

TURNS BACK INTO LAVA

Their original structure is changed by heat and pressure.

TURNS BACK INTO LAVA

② **IGNEOUS ROCK**

Rock formed when magma (or lava at Earth's surface) solidifies. Basalt and granite are good examples of igneous rocks.

① **LAVA**

The name for magma emerging as a liquid on Earth's surface.

## SOLID LAVA

Lava solidifies at temperatures below 900° C (1,900° F). The most viscous type of lava forms a rough landscape, littered with sharp rocks; more fluid lava, however, tends to form flatter and smoother rocks.

# 1,000° C (1,800° F)

**IS THE AVERAGE TEMPERATURE OF LIQUID LAVA.**

## TYPES OF LAVA

Basaltic lava is found mainly in islands and in mid-ocean ridges; it is so fluid that it tends to spread as it flows. Andesitic lava forms layers that can be up to 130 feet (40 m) thick and that flow very slowly, whereas rhyolitic lava is so viscous that it forms solid fragments before reaching the surface.

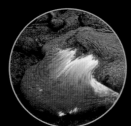

**ANDESITIC LAVA**

Silicates
63%

Other Content
37%

**RHYOLITIC LAVA**

Silicates
68%

Other Content
32%

# MINERALS: THE "BRICKS"

**Minerals are the "bricks" of materials that make up the Earth and all other solid bodies in the Universe.** They are usually defined both by their chemical composition and by their orderly internal structure. Most are solid crystalline substances. However, some minerals have a disordered internal structure and are simply amorphous solids similar to glass. Studying minerals helps us to understand the origin of the Earth. Minerals are classified according to their composition and internal structure, as well as by the properties of hardness, weight, color, luster, and transparency. Although more than 4,000 minerals have been discovered, only about 30 are common on the Earth's surface.

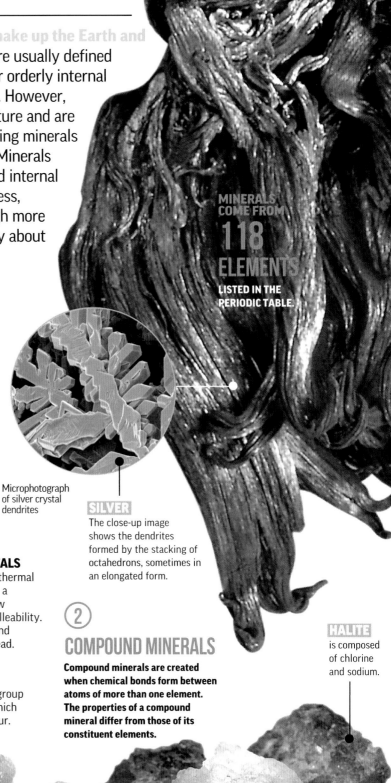

**MINERALS COME FROM 118 ELEMENTS LISTED IN THE PERIODIC TABLE.**

Microphotograph of silver crystal dendrites

## COMPONENTS

The basic components of minerals are the chemical elements listed on the periodic table. Minerals are classified as native if they are found in isolation, contain only one element, and occur in their purest state. On the other hand, they are classified as compound if they are composed of two or more elements. Most minerals fall into the compound category.

### ① NATIVE MINERALS

**These minerals are classified into:**

**GOLD**
An excellent thermal and electrical conductor. Acids have little or no effect on it.

**SILVER**
The close-up image shows the dendrites formed by the stacking of octahedrons, sometimes in an elongated form.

**METALS AND INTERMETALS**
Native minerals have high thermal and electrical conductivity, a typically metallic luster, low hardness, ductility, and malleability. They are easy to identify and include gold, copper, and lead.

**SEMIMETALS**
Native minerals that are more fragile than metals and have a lower conductivity. Examples are arsenic, antimony, and bismuth.

**NONMETALS**
An important group of minerals, which includes sulphur.

### ② COMPOUND MINERALS

**Compound minerals are created when chemical bonds form between atoms of more than one element. The properties of a compound mineral differ from those of its constituent elements.**

**HALITE**
is composed of chlorine and sodium.

BISMUTH

SULPHUR

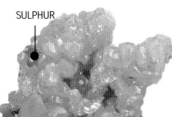

## POLYMORPHISM

A phenomenon in which the same chemical composition can create multiple structures and, consequently, result in the creation of several different minerals. The transition of one polymorphous variant into another, facilitated by temperature or pressure conditions, can be fast or slow and either reversible or irreversible.

| Chemical Composition | Crystallization System | | Mineral |
|---|---|---|---|
| $CaCO_3$ | | Trigonal | Calcite |
| $CaCO_3$ | | Rhombic | Aragonite |
| $FeS_2$ | | Cubic | Pyrite |
| $FeS_2$ | | Rhombic | Marcasite |
| C | | Cubic | Diamond |
| C | | Hexagonal | Graphite |

### DIAMOND AND GRAPHITE

A mineral's internal structure influences its hardness. Both graphite and diamond are composed only of carbon; however, they have different degrees of hardness.

Diamond

Graphite

**MORE THAN**

# 4,000
## TYPES OF MINERALS

**HAVE BEEN RECOGNIZED BY THE INTERNATIONAL ASSOCIATION OF MINERALOGY.**

## ISOTYPIC MINERALS

Isomorphism happens when minerals with the same structure, such as halite and galena, exchange cations. The structure remains the same, but the resulting substance is different, because one ion has been exchanged for another. An example of this process is siderite (rhombic $FeCO_3$), which gradually changes to magnesite ($MgCO_3$) when it trades its iron (Fe) for similarly-sized magnesium (Mg). Because the ions are the same size, the structure remains unchanged.

Carbon Atom

Model demonstrating how one atom bonds to the other four

Each atom is joined to four other atoms of the same type. The carbon network extends in three dimensions by means of strong covalent bonds. This provides the mineral with an almost unbreakable hardness.

**Hardness of 10**
on the Mohs scale.

Atoms form hexagons that are strongly interconnected in parallel sheets. This structure allows the sheets to slide over one another.

**Hardness of 1**
on the Mohs scale.

### HALITE AND GALENA

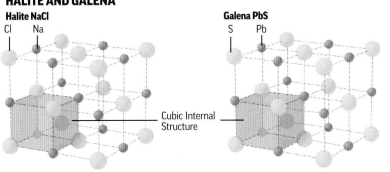

**Halite NaCl**

Cl    Na

**Galena PbS**

S    Pb

Cubic Internal Structure

# THE ORIGIN OF LIFE

Life began on Earth approximately 3.8 billion years ago in the form of microbes, which determined, and continue to determine, biological processes on the planet. Science tries to explain the source of life through a series of chemical reactions, which occurred by chance. Over the course of millions of years, they would give rise to different living organisms.

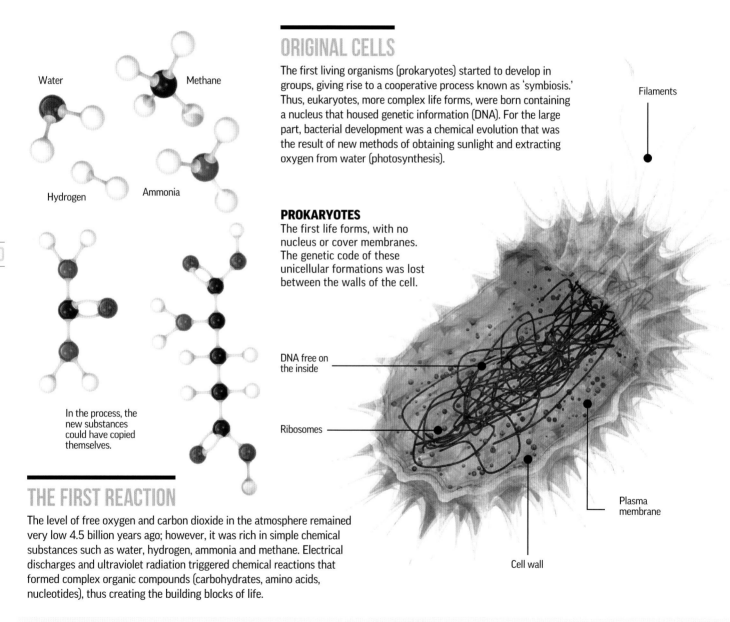

Water

Methane

Hydrogen

Ammonia

In the process, the new substances could have copied themselves.

## THE FIRST REACTION

The level of free oxygen and carbon dioxide in the atmosphere remained very low 4.5 billion years ago; however, it was rich in simple chemical substances such as water, hydrogen, ammonia and methane. Electrical discharges and ultraviolet radiation triggered chemical reactions that formed complex organic compounds (carbohydrates, amino acids, nucleotides), thus creating the building blocks of life.

## ORIGINAL CELLS

The first living organisms (prokaryotes) started to develop in groups, giving rise to a cooperative process known as 'symbiosis.' Thus, eukaryotes, more complex life forms, were born containing a nucleus that housed genetic information (DNA). For the large part, bacterial development was a chemical evolution that was the result of new methods of obtaining sunlight and extracting oxygen from water (photosynthesis).

**PROKARYOTES**
The first life forms, with no nucleus or cover membranes. The genetic code of these unicellular formations was lost between the walls of the cell.

Filaments

DNA free on the inside

Ribosomes

Plasma membrane

Cell wall

## ARCHAEAN
### 4.8 billion years ago

The atmosphere distinguishes Earth from the other planets.

### 4.5 billion years ago

Volcanic eruptions and igneous rock dominate the Earth's landscape.

### 4.2 billion years ago

The Earth's surface cools and accumulates liquid water.

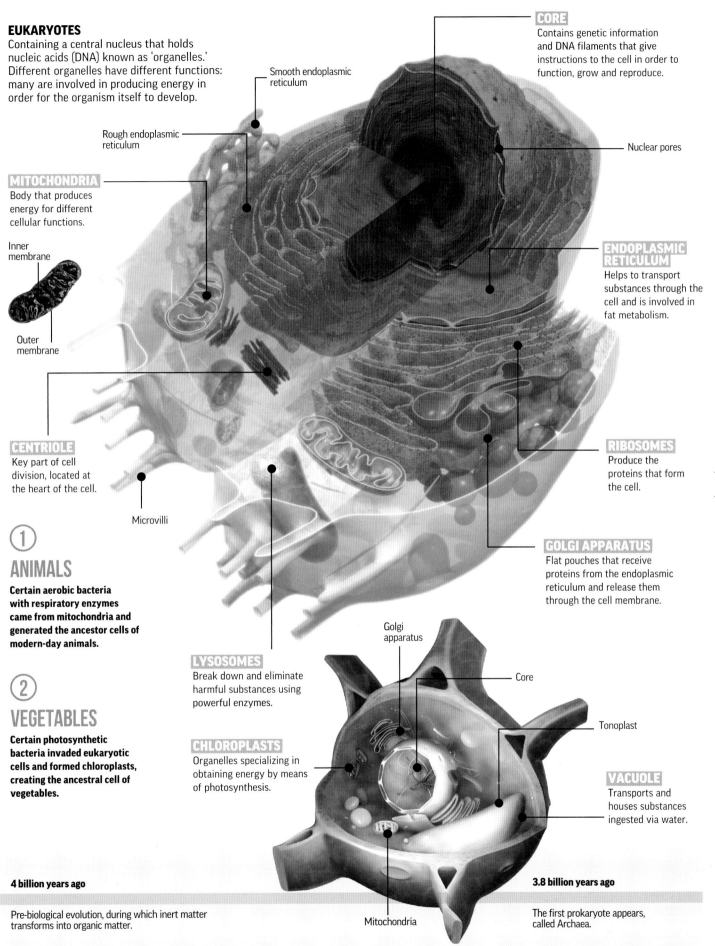

## EUKARYOTES

Containing a central nucleus that holds nucleic acids (DNA) known as 'organelles.' Different organelles have different functions: many are involved in producing energy in order for the organism itself to develop.

Smooth endoplasmic reticulum

Rough endoplasmic reticulum

### CORE
Contains genetic information and DNA filaments that give instructions to the cell in order to function, grow and reproduce.

Nuclear pores

### MITOCHONDRIA
Body that produces energy for different cellular functions.

Inner membrane

Outer membrane

### ENDOPLASMIC RETICULUM
Helps to transport substances through the cell and is involved in fat metabolism.

### CENTRIOLE
Key part of cell division, located at the heart of the cell.

Microvilli

### RIBOSOMES
Produce the proteins that form the cell.

### GOLGI APPARATUS
Flat pouches that receive proteins from the endoplasmic reticulum and release them through the cell membrane.

① 
## ANIMALS
**Certain aerobic bacteria with respiratory enzymes came from mitochondria and generated the ancestor cells of modern-day animals.**

② 
## VEGETABLES
**Certain photosynthetic bacteria invaded eukaryotic cells and formed chloroplasts, creating the ancestral cell of vegetables.**

### LYSOSOMES
Break down and eliminate harmful substances using powerful enzymes.

### CHLOROPLASTS
Organelles specializing in obtaining energy by means of photosynthesis.

Golgi apparatus

Core

Tonoplast

### VACUOLE
Transports and houses substances ingested via water.

Mitochondria

**4 billion years ago**

Pre-biological evolution, during which inert matter transforms into organic matter.

**3.8 billion years ago**

The first prokaryote appears, called Archaea.

# FOSSIL RELICS

**The oldest fossils found thus far date back to the end of the Proterozoic Period, during the Precambrian era.** Found in Ediacara (Australia), they represent the first evidence of multicellular organisms with different tissues. It is believed that these specimens were not animals, but prokaryote organisms formed by several cells with no internal cavities.

## FIRST SPECIES

It has been established that the Ediacara organisms were the first invertebrates on Earth. They appeared approximately 650 million years ago and were made up of several cells. Some had a soft, flat body while others were shaped like a disc, or a long strip. A single cell was no longer responsible for feeding itself, breathing and reproducing; instead, several cells specialized in different functions.

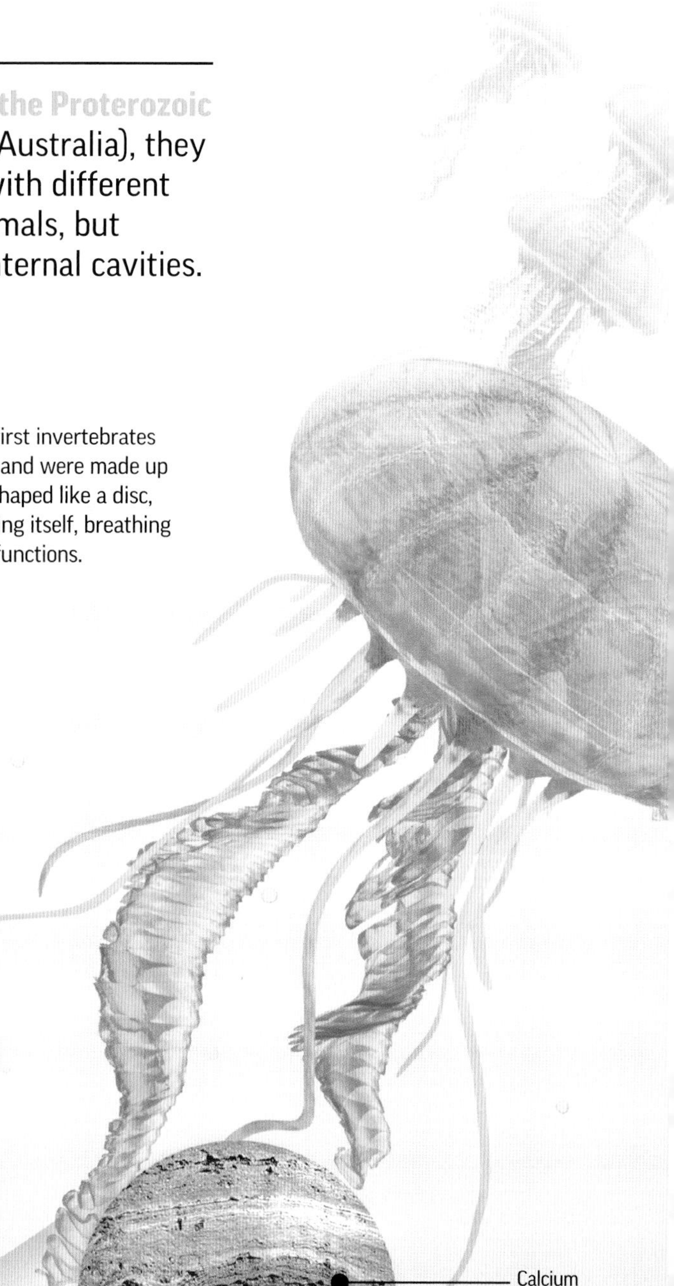

**CHARNIA**

One of the largest fossils found in Ediacara. Its flattened, leaf-shaped body was attached to a structure similar to a disc that attracted organisms to the bottom of the sea.

## 100 M
### (328 FT)

**ITS MAXIMUM LENGTH.**

**STROMATOLITES**

They provide the most ancient records of life on Earth. They were layered structures, primarily cyanobacterias and calcium carbonate, that stuck to the substrate. They grew in mass, helping the formation of reefs.

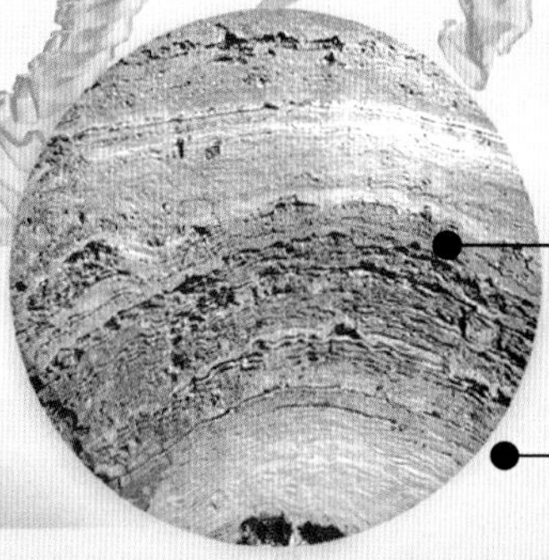

Calcium carbonate

Cynobacteria

| 3.5 billion years ago | 2.5 billion years ago | 650 million years ago |
|---|---|---|
| Accumulation of iron oxide on the seabed. | Formation of stromatolite reefs. | Traces of Ediacara fauna, amongst the oldest known. |

## MAWSONITES

This type of jellyfish moved slowly through the water, assisted by the currents. It contracted its umbrella, extended its tentacles and shot its microscopic spears to capture its prey.

**9-10 CM**
**(3.5-4 IN)**
**IN DIAMETER.**

## CYCLOMEDUSA

Ancient circular fossil with a bump in the middle and up to five concentric ridges. Some radial segment lines extended across the outer discs.

## KIMBERELLA

The first known organism to contain a body cavity. It is believed to have been similar in size to a mollusc.

**20 CM**
**(7.9 IN)**
**IN LENGTH.**

**3 CM**
**(1.2 IN)**
**IN LENGTH.**

## TRIBRACHIDIUM

It is believed that this species, which developed in the shape of a disc with three symmetrical parts, is a distant relative of corals, anemones and starfish.

## DICKINSONIA

Often considered a ringed worm given its similarity to an extinct species (Spinther). It may also have been a version of the soft bodied fungus, 'banana coral.'

**5 CM**
**(2 IN)**
**IN DIAMETER.**

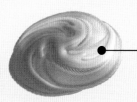

**542 million years ago**

The Cambrian Period begins. Significant development in multicellular life forms.

**540 million years ago**

The period to when the first remnants of invertebrates date, found in Canada's Burgess Shale.

# CAMBRIAN EXPLOSION

**The great explosion of life that occurred during the Cambrian Period, around 500 million years ago,** gave rise to a wide variety of multicellular organisms protected by exoskeletons or shells. However, although these organisms represent the fauna typical of the Cambrian Period, several species of soft animals lived side-by-side with them during the same period.

## THE BURGESS SHALE FIELD

Located in the Yoho National Park, in the Canadian province of British Columbia, the Burgess Shale is a famous fossil field discovered in 1909 by US palaeontologist Charles Walcott. The Burgess Shale is the world's greatest reservoir of soft animal fossils from the Cambrian period. It is home to thousands of extremely well-preserved fossilized invertebrates, such as arthropods, worms and primitive chordates.

10 mm (0.4 in)

Equipped with a powerful exoskeleton, the Anomalocaris truly terrorized the Cambrian seas.

**SPONGES**
They often developed together with algae of different species, sizes and shapes.

**PRIAPULIDA**
Benthic worms that lived in the sand and in the slush of deep or shallow waters.

**2 CM**
**(0.8 in)**
**IN LENGTH.**

**CAMBRIAN**
**(534 to 490 million years ago)**

Beginnings of the Cambrian Period

The increased amount of oxygen helped in the formation of shells.

### ANOMALOCARIS
The largest predator arthropod known at the time. It had a circular mouth, appendages that allowed it to firmly grasp its prey and fins along both sides for swimming.

**HUMAN SCALE COMPARISON**
# 60 CM
## (24 IN)
**IN LENGTH.**

### PIKAIA
One of the first chordata, similar to swimming worms, with a tail in the shape of a fin. It is the oldest known ancestor of vertebrates.

# 10 CM
## (4 IN)
**IN LENGTH, INCLUDING ITS TAIL.**

### MARRELLA
A tiny swimming arthropod, probably easy prey to the Burgess Shale predators.

# 10 CM
## (4 IN)
**IN LENGTH, UP TO ITS EXTREMITIES.**

### HALLUCIGENIA
The defense system of this arthropod was its long spikes which served as legs.

# 10 CM
## (4 IN)
**MAXIMUM LENGTH.**

**Evolutionary explosion**

The Cambrian Period saw the formation of a wide range of body designs.

**Coral reefs**

Comprising countless soft-bodied animals.

# CONQUEST OF THE EARTH

The Paleozoic Era (ancient life) was witness to a turning point in the evolutionary process: the conquest of land around 360 million years ago. For this process, diverse mechanisms of adaptation were necessary, from new designs of vascular plants and changes in the bone and muscular structures to new systems of reproduction. The appearance of reptiles and their novel amniotic egg meant the definitive colonization of the land by the vertebrates, just as the pollen made plants completely independent of water.

## 6 MM (0.2 IN)
**THE APPROXIMATE WINGSPAN OF THE *MEGANEURA*.**

**JAW**
Key to the evolution of vertebrates, allowing them to become predators.

## NEW FISH SPECIES

The period saw the development of armored, jawless fish, the first known vertebrates. Jawed and freshwater fish would appear later; their evolution coincided with the predominance of bony fish, from which amphibians evolved.

## 500 CM (197 IN)
**THE MAXIMUM LENGTH OF THE *DUNKLEOSTEUS*.**

Barracuda skull

Dorsal fin

Thin, lobed fin

Head and chest armor connected

Bony teeth with sharp edges

The Devonian Period is known as the age of fish.

**FIN**
To move through the water, the Acanthostega moved its fin from side to side. It kept this fin during its transition to move onto land.

**DUNKLEOSTEUS**
Human scale comparison.

| ORDOVICIAN | SILURIAN | DEVONIAN |
|---|---|---|
| 490 to 443 million years ago | 443 to 418 million years ago | 418 to 354 million years ago |
| The first organisms – lichens and bryophytes – appear. | Great coral reefs and certain types of small plants. | Vascular plants and arthropods form a number of ecosystems on dry land. |

# FROM FINS TO EXTREMITIES

Amphibian evolution facilitated the exploration of new food sources, such as insects and plants, and the adaptation to a new respiratory system. Aquatic vertebrates therefore had to modify their skeleton and develop muscle. Meanwhile, their fins were turned into legs, which allowed them to move over land.

**ACANTHOSTEGA**
Human scale comparison.

## 90-120 CM
### (35-47 IN)
**MAXIMUM LENGTH.**

**BACKBONE**
With a system of projections, known as 'zygapophyses,' between the vertebrae, they were able to keep their backbone rigid.

**PREDATOR**
It developed a large mouth which allowed it to hunt other vertebrates.

**BONE STRUCTURE**
Just three bones (humerus, radius and ulna) formed the bone composition of its leg. Unlike fish, they had a movable wrist and eight fingers that moved together like a spade.

**AMNIOTIC EGG**
The success of vertebrates' colonization of land is attributed to the evolution of the amniotic egg, enveloped by a leathery cover.

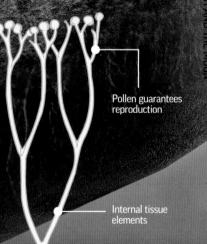

Air chamber
Albumen
Shell
Yolk sac
Chorion
Embryo
Amnion
Allantois

## 6 MM
### (0.2 IN)

**VASCULAR DEVELOPMENT OF PLANTS**
The need to transport water from the root to the stalk and the products of photosynthesis in the opposite direction resulted in plants developing a system of internal tissues. Pollen-based reproduction allowed plants to definitively adapt on land.

Pollen guarantees reproduction

Internal tissue elements

**CARBONIFEROUS**
**354 to 290 million years ago**

Land tetrapods and insects with wings appear.

**PERMIAN**
**290 to 252 million years ago**

A wide variety of insects and vertebrates appear on land.

# THE TREE OF LIFE

**This image, a phylogenetic tree, explains how all living beings are related.** It is compiled using information from fossils and data obtained from a structural and molecular comparison of organisms. Phylogenetic trees are based on the theory that all organisms descend from a common ancestor: the protocell.

## EUKARYOTE

Encompasses living species whose cellular make-up contains a genuine nucleus. Covers unicellular and multicellular organisms.

## ANIMALS

Multicellular and heterotrophs. Characterized by their mobility and internal organ systems. They sexually reproduce and their metabolism is aerobic.

## ARCHAEA

These organisms are unicellular and microscopic. Most are anaerobic and live in extreme environments, half release methane as part of their metabolic process. There are 502 known species.

## PLANTS

Multicellular autotrophic organisms containing cells with a nucleus. They perform photosynthesis using chloroplasts.

**Cnidarians**
Include species like jellyfish, polyps and corals.

**Bilaterality**
Bilaterally symmetrical organisms.

**Euryarchaeota**
*Halobacterium salinarum.*

**Korarchaeota**
This group is amongst the most primitive.

**Non-vascular plants**
With no internal vascular system.

**Vascular plants**
With an internal vascular system.

**Molluscs**
Include octopuses, snails and oysters.

**Vertebrates**
They have a backbone, a skull that protects their brain and a skeleton.

**Crenarchaeota**
Found at high temperatures.

**Seed-bearing plants**
Plants on which the seed is exposed and bear flowers or fruits.

**Tetrapods**
Comprising four members.

**Non-seed-bearing plants**
Featuring simple tissue and stalks on which the surface takes precedence.

**Cartilaginous fish**
Include rays, manta rays and sharks.

## ORIGINS

Available scientific evidence sustains that all species share a common ancestry. However, there is no conclusive data regarding their origin. It is known that the first expression of life must have been a prokaryote – unicellular beings whose genetic information was contained anywhere within the cell walls.

**Gymnosperm**
On which the seeds are naked.

**Angiospermae**
Flowering, fruit-bearing plants. Encompassing more than 250,000 species.

**Amphibians**
Born in water and live on land.

# BACTERIA

Unicellular organisms that live on surfaces in colonies. Generally, they have a cell wall made up of peptidoglycans and many have cilia. It is believed that they have existed for 3.8 billion years.

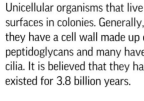

**Cocci**
Pneumococci are an example.

**Bacillus**
*E. coli* takes on this form.

**Spirochaetes**
Helically or spirally shaped.

**Vibrio**
Found in salt water.

## PROTISTS

Includes species that cannot be classified within other groups, such as algae and amoebae.

## FUNGI

Heterotrophic cellular organisms with cellular walls thickened by chitin. Use external digestion.

**Basidiomycetes**
Include button mushrooms.

**Zygomycetes**
They reproduce by means of zygosporangia.

**Ascomycetes**
The highest number of species fall into this group.

**Chytridiomycetes**
Some even have mobile cells.

**Deuteromycetes**
Not known to reproduce sexually.

# 10,000,000
**SPECIES OF ANIMALS ARE CALCULATED TO INHABIT EARTH IN ITS DIFFERENT ENVIRONMENTS.**

**Arthropods**
They have an outer skeleton (exoskeleton). Their extremities are articulated appendages.

**Insects**
In evolutionary terms, the most successful.

**Myriapoda**
Millipedes and centipedes.

**Bony fish**
Equipped with bones and a jaw.

**Crustaceans**
Crabs and lobsters.

**Arachnids**
Spiders, scorpions and mites.

## AMNIOTES

Species in this group protect their embryo in a sealed structure – the amniotic egg. Only some mammals are still oviparous. However, of placental mammals (like humans), the placenta is a modified egg; its membranes have been transformed, but the embryo continues to be surrounded by an amnion full of amniotic fluid.

**Amniotes**
Species born from an embryo within an amniotic egg.

**Mammals**
Their young feed on their mother's milk.

**Placental mammals**
Their young are born in a fully developed state.

## MAN

Human beings form part of the order of Primates, one of the 19 orders into which placental or eutheria mammals are placed. We share characteristics with monkeys and apes and our closest relatives are great apes, like the chimpanzee or gorilla.

**Birds and reptiles**
Oviparous species.

**Marsupials**
The embryo completes its development outside the womb.

**Monotremes**
The only oviparous mammals. They are the most primitive form of mammals.

**Tortoise**
The oldest reptile.

**Lizards**
Including crocodiles.

**Snakes**
Squamata and long-bodied.

# 5,416
**SPECIES OF MAMMALS INCLUDED IN THE THREE GROUPS.**

# OTHER ORIGIN?

**Some geophysicists maintain that nearly 40 billion years ago Earth did not contain the conditions necessary to sustain organic matter.** An alternative theory is that the first complex molecules and simple organic compounds arrived on Earth from outer space, embedded in comets and asteroids.

## PANSPERMIA

This theory dates back to Ancient Greece, and argues that life-bearing germs exist all over the Universe. Consequently, life on Earth would not exist had it not been propagated from germs arriving on comets and meteoroids. This theory is founded on studies which conclude that bacterial spores could survive the extreme conditions of outer space, including voyages throughout the galaxy, for thousands of years.

### SVANTE ARRHENIUS

This Swedish scientist, the 1903 Nobel Prize in Chemistry winner, was a huge proponent of panspermia. The theory—which holds that the "seed" of life traveled through solar radiation—was invalidated by Paul Becquerel, who demonstrated that microorganisms cannot survive ultraviolet rays.

### POSSIBLE ORIGINS

If life did not originate on Earth, then where did it come from? Astrobiology points out planets and satellites where water may have existed during the embryonic phase of the Solar System.

1. **Mars.** The red planet is responsible for meteorites ejected from its orbit that finally landed on Earth.

2. **Europa.** Jupiter's sixth satellite in terms of distance has an icy surface and possible water interior where microorganisms may exist.

3. **Titan.** Saturn's largest satellite is rich in organic compounds.

4. **Venus.** Although extremely hot and with excessive atmospheric pressure, at one time it may have sustained microbial life in its atmosphere.

## 5% OF METEORITES

**THAT EJECTED FROM MARS EARLY IN THE FORMATION OF THE SOLAR SYSTEM MAY HAVE IMPACTED EARTH.**

## AGAINST

Critics of panspermia point out that no bacteria can survive the heat of an atmospheric entry and the force of impact of a meteor. Also, this theory only transfers the mysteries surrounding the origins of life to another source in the Cosmos.

## LITHOPANSPERMIA

In the 1960s, physicists Fred Hoyle and Chandra Wickramasinghe were influential proponents of this variation of panspermia—the belief that the origins of life on Earth originated from the transfer of organisms in rocks from one planet to another that impacted Earth millions of years ago. In 1996, research conducted on the Mars meteor Allan Hills 84001 claimed to have found possible evidence of microscopic fossils of Martian bacteria. This has become the principal proof of lithopanspermia.

## MURCHISON

This meteor fell to Earth near Murchison, Victoria in Australia in 1969 and contains millions of organic compounds. Evidence that lithopanspermia exists.

### ALLAN HILLS 84001

A paper published by geologist David McKay, chief scientist for astrobiology at NASA, confirms the presence of fossilized bacteria containing remains of microscopic life forms.

## DIRECTED PANSPERMIA

A less scientific and more speculative hypothesis proposes the deliberate or accidental transport of microorganisms by other intelligent extraterrestrial life forms to introduce species to Earth.

**MODERN OBSERVATORY**
The antennas of the Atacama Large
Millimeter/submillimeter Array (ALMA)
stand out against the Chilean night sky.

# CHAPTER 4

# HISTORY OF
# ASTRONOMY

Astronomy was born out of humankind's need to measure time and seasons, marking the best times to plant. In ancient times, the study of the stars was mixed with superstition and ritual. Today, thanks to advances in new technologies, such as the giant telescopes installed in various locations around the planet, we have discovered many new things about the universe.

# THE FIRST ASTRONOMERS

**Since the dawn of humanity, civilizations have developed their own ideas about the stars.** They sought gods, divine messages, signs of their prophecies and points of reference for their calendars in the heavens. It can be asserted that the origins of astronomy can be found in the careful and systematic observation of the stars undertaken by these ancient peoples to satisfy their own social needs, such as measuring time or planning their agriculture.

## CALENDARS AND·FORECASTS

In short, astronomy was born from necessity. Over 5,000 years ago, the Egyptians developed a calendar based on the solar cycle (very similar to the one we use today), which helped them to identify dates for sowing, cultivation, harvesting and flooding the Nile Delta, where they lived. In turn, Mesopotamian populations (Sumerians and Babylonians) depended on climate variations for their subsistence, inventing astrology as a way of 'forecasting' and controlling their environment.

### SUMERIAN ASTROLOGY

In ancient Mesopotamia, astronomy (a natural science) and astrology (divination using the stars) formed part of the same set of knowledge. In Western societies, these fields were separated after the Renaissance.

### SCRIBES

Scribes, or sangha, preserved religious, literary and scientific knowledge. Centuries later, the first Sumerian pictograms traced wedge-shaped lines.

### SLATS

Clay fragments used to record the position in which the Sun rose and set, the stellar groups and the phases of the Moon.

## MAYAN ASTRONOMY

The astronomic knowledge of Mesoamerican civilizations was surprisingly advanced. Among them, the Mayans are particularly noteworthy; their priests understood the movement of the stars and could predict eclipses and the path of Venus. Their calendar was also extremely accurate.

### MAYAN WISDOM

Mayan constructions reflected their astronomic knowledge. The temple at Chichén Itzá, for example, has 365 steps; the same number as the days of the year.

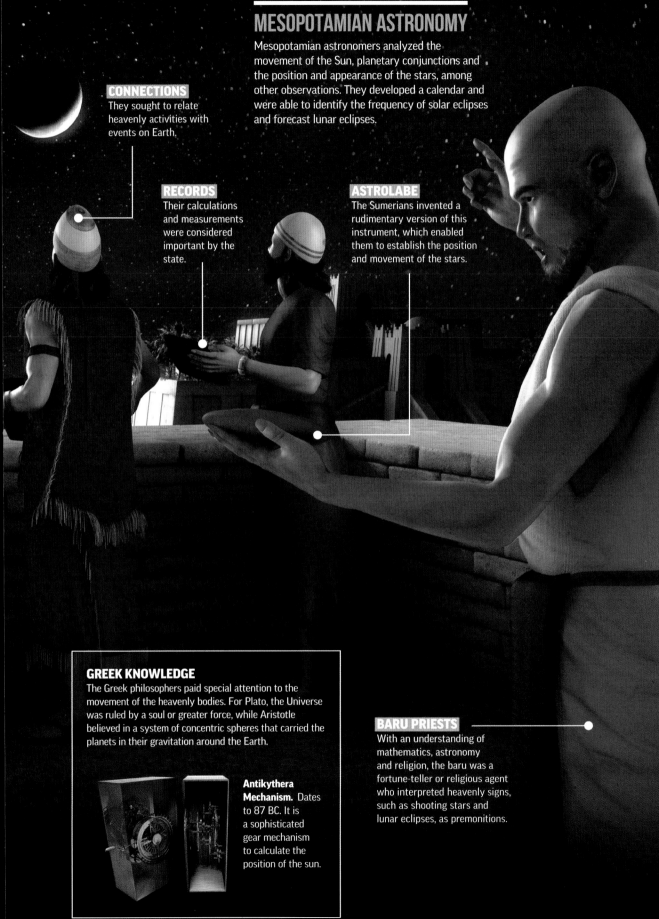

# MESOPOTAMIAN ASTRONOMY

Mesopotamian astronomers analyzed the movement of the Sun, planetary conjunctions and the position and appearance of the stars, among other observations. They developed a calendar and were able to identify the frequency of solar eclipses and forecast lunar eclipses.

## CONNECTIONS
They sought to relate heavenly activities with events on Earth.

## RECORDS
Their calculations and measurements were considered important by the state.

## ASTROLABE
The Sumerians invented a rudimentary version of this instrument, which enabled them to establish the position and movement of the stars.

## GREEK KNOWLEDGE
The Greek philosophers paid special attention to the movement of the heavenly bodies. For Plato, the Universe was ruled by a soul or greater force, while Aristotle believed in a system of concentric spheres that carried the planets in their gravitation around the Earth.

**Antikythera Mechanism.** Dates to 87 BC. It is a sophisticated gear mechanism to calculate the position of the sun.

## BARU PRIESTS
With an understanding of mathematics, astronomy and religion, the baru was a fortune-teller or religious agent who interpreted heavenly signs, such as shooting stars and lunar eclipses, as premonitions.

# ASTRONOMICAL THEORIES

For a long time, it was believed that the Earth did not move, that it was static and that the Sun, the Moon and the planets orbited around it. With the invention of the telescope, our way of looking at the Universe changed. The center was no longer occupied by the blue planet, and the concept that all planets orbited the Sun was established.

## GEOCENTRIC MODEL

The great promoter of the geocentric model (according to which the Earth was the center of the Universe) was Egyptian astronomer Claudius Ptolemy; in the second century ce, he collated the astronomical ideas of the ancient Greeks, Aristotle in particular. Although other ancient astronomers, such as Aristarchus of Samos, asserted that the Earth was round and that it orbited the Sun, Aristotle's proposals were adopted, preserved and defended by the Catholic Church as the commonly accepted beliefs for 16 centuries.

### MEASUREMENTS

Having observed that the Sun, the Moon and the stars moved cyclically, ancient civilizations discovered that they could use the heavens as a clock and a calendar. However, they found it difficult to simplify complex calculations when forecasting the position of the stars. A tool that would serve them to this end was the astrolabe.

#### ASTROLABE
The different engraved plates reproduce the celestial spheres in two dimensions, so that the height of the Sun and the stars can be measured.

### GREAT ASTRONOMERS

**2ND CENTURY**
**Claudius Ptolemy (90–168)**
He collated the work of renowned Greek astronomers. He became the unopposed leader of the field.

**16TH CENTURY**
**Nicolaus Copernicus (1473–1543)**
He proposed that the Sun was the center of the Universe, not the Earth.

**17TH CENTURY**
**Johannes Kepler (1571–1630)**
German astronomer who formulated three renowned laws of planetary motion.

## HELIOCENTRIC MODEL

In 1543, just a few months before his death, Nicolaus Copernicus published his book, *De revolutionibus orbium coelestium*, the result of which would become what is now know as the 'Copernican Revolution.' The Polish astronomer developed the heliocentric theory, which opposed the geocentric theory. Heliocentrism established the Sun as the center of the Universe and the Earth as just another satellite. This theory contrasted with the teachings of the Church, with both the Catholic and Protestant churches banning all documents that promoted these beliefs. Galileo Galilei himself was prosecuted for sustaining this undeniable truth.

### GALILEO'S TELESCOPE

It would appear that the telescope was invented in 1608 by Dutch lens-maker Hans Lippershey. However, it served no scientific purpose until Galileo Galilei developed it for use in observing the heavens.

### FUNCTIONALITY

Galileo's first telescope was a leather tube with two lenses facing each other on both ends – one was convex and the other was concave.

## 30 TIMES

**THE MAGNIFICATION OF OBJECTS USING THE FIRST TELESCOPE BUILT BY GALILEO.**

### 17TH CENTURY
**Galileo Galilei (1564–1642)**
He discovered solar spots, the four moons of Jupiter, the phases of Venus and the Moon's craters.

### 17TH CENTURY
**Isaac Newton (1643-1727)**
He developed the theory of gravitation: all the heavenly bodies are governed by the same laws.

### 20TH CENTURY
**Edwin Hubble (1889-1953)**
He investigated the expansion of the galaxies, enabling the formulation of the Big Bang theory.

# SPRINKLED WITH STARS

The constellations take the names of different animals, mythical characters or other figures assigned to them by ancient civilizations to use them as points of reference or to guide them on their travels. They are groups of stars that appear aligned, whereas in reality, there are significant distances between them. Depending on the hemisphere from which they are observed, the hour and time of the year, different constellations are visible.

### ORIGINS

The origin of the constellations in Western culture can be traced to the first astronomical observations of the Mesopotamian peoples: the Greco-Latin culture preserved them and as a result, most have Classic mythical names. Later, as other constellations were discovered, they were most commonly named after phenomena related to science.

### THE CHANGING SKIES

As the Earth follows its orbit, the sky changes, with different areas of space coming into view. Therefore, the constellations that can be seen depend on the time of year and the latitude from which they are observed. It is only from the equator that all constellations can be seen in a single night.

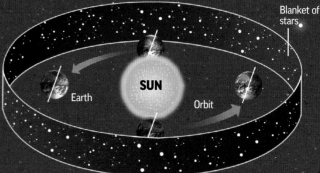

Blanket of stars

SUN

Earth

Orbit

## 88
### CONSTELLATIONS
**ARE RECOGNIZED BY THE INTERNATIONAL ASTRONOMICAL UNION.**

Chi 1 Orionis

Xi Orionis

Mu Orionis

Betelgeuse

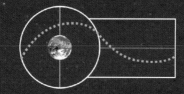

**LEO**
The brightest star is Regulus, which is the lion's heart.

**CANCER**
The least spectacular of the 13.

**GEMINI**
The stars Castor and Pollux appear as the heads of the twins.

**TAURUS**
The brightest star, Aldebaran, is red.

**ARIES**
It has just one very bright star: Hamal ('sheep' in Arabic).

**PISCES**
It doesn't have any bright stars.

## THE ZODIAC

The 13 constellations positioned within the elliptical path of Earth, where the Sun appears to travel as seen from Earth, are called the 'zodiac constellations' and form the basis of astrology. However, Ophiuchus, the 13th, is not considered by astrologers.

### NORTH AND SOUTH
The zodiac constellations can be seen from both hemispheres. However, it is difficult to see Scorpio from the north; whereas it is difficult to see constellations such as Gemini from the south.

## ORION

Its name alludes to the mythical Greek character: a giant and handsome hunter. It is one of the most commonly known constellations, as it can be seen from both hemispheres throughout the entire night.

Omicron Orionis

Pi 1 Orionis

Pi2 Orionis

Pi3 Orionis

Pi4 Orionis

Heka

Bellatrix

Pi5 Orionis

Pi6 Orionis

### THE MYSTERY OF GIZA
The three stars in Orion's belt appear closely related to the alignment of the three pyramids in Egypt.

Mintaka

Alnilam

Alnitak

Saiph

Rigel

### OPHIUCHUS
It is not recognized within the zodiac, because when astrology was born around 3,000 years ago, it was too far from the elliptical path of the Earth.

### DIFFERENT CULTURES

In ancient times, each culture developed constellations that were very different to one another. For example, the Chinese used smaller, more detailed patterns, making it possible to obtain more accurate information about positions. Thus, different names may be given to the same constellations.

### SCORPIO
Known to Mesopotamian peoples, Greece, Rome, Mesoamerica and Oceania.

### URSA MAJOR
The bear used to depict this constellation is strange, as it has a large tail.

### CENTAURUS
A half-man, half-horse creature from Greek mythology. He accompanied Orion in the quest to recover his sight.

### BABYLONIA

The Babylonians were responsible for giving roots to the concept of the zodiac 2,000 years BCE as a way of measuring time. As such, it served as a symbolic calendar.

### LIBRA
It was once part of Scorpio.

### SAGITTARIUS
At the center of the Milky Way, full of nebulae and groups of stars.

### VIRGO
The constellation with the brightest stars.

### AQUARIUS
Contains globular clusters and nebulae.

### CAPRICORN
One of the least impressive.

### SCORPIO
It points towards the Milky Way. Its brightest star is Antares.

# THE ASTRONOMICAL CLOCK OF SU SONG

**Between the fourth century BCE and the thirteenth century CE, China experienced extraordinary scientific and technical advances, the results of which would remain unknown to the West for a long time.** Among these events, the sensational, giant astronomical clock created by Su Song is particularly noteworthy. Invented in 1088, it was the first highly accurate clock and it demonstrates the advanced knowledge of astronomy held by the Chinese civilization in the Middle Ages.

## PRECISION IN THE FORM OF A CLOCK

Although scarce evidence of his ingenuity remains, Su Song's treatise about the clock tower, Xinyi Xiangfayao, was published in 1092. Construction of the clock was completed in 1090, after confirming the functionality of a prototype. Powered using water, its inaccuracy was less than 100 seconds per day and showed the stellar constellations independently of weather conditions.

**WORKERS**
Given its size, the tower was capable of housing several operators.

**ARMILLARY SPHERE**
Invented in 978 and comprising a series of rings, it was the first instrument to determine the position of the stars until the invention of the telescope in the seventeenth century.

**MAIN WHEEL**
It measured 3 m (10 ft) in diameter and contained 36 vanes. Its system of cogs transferred the required power to activate the time and the armillary sphere mechanisms.

**MATERIAL**
The tower was made from wood. Its most important components were cast in bronze.

**SU SONG'S CONTRIBUTIONS**
Su Song (1020–1101) had many occupations, including civil servant, engineer, botanist, poet, antiquarian and ambassador of the Song Dynasty. Although his most famous work was the Great Clock Tower, he made numerous contributions to other scientific fields. In addition to treatises on mineralogy, botany and pharmacology, he produced maps of the Earth with time zones, and an atlas with several maps of the stars.

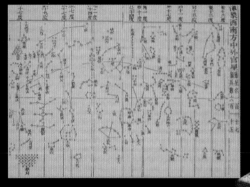

**MAP OF THE STARS**
Published by Su Song in a treatise in 1092, it is one of the oldest printed star charts in the world.

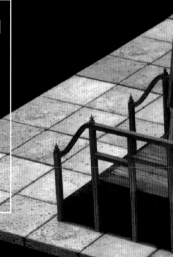

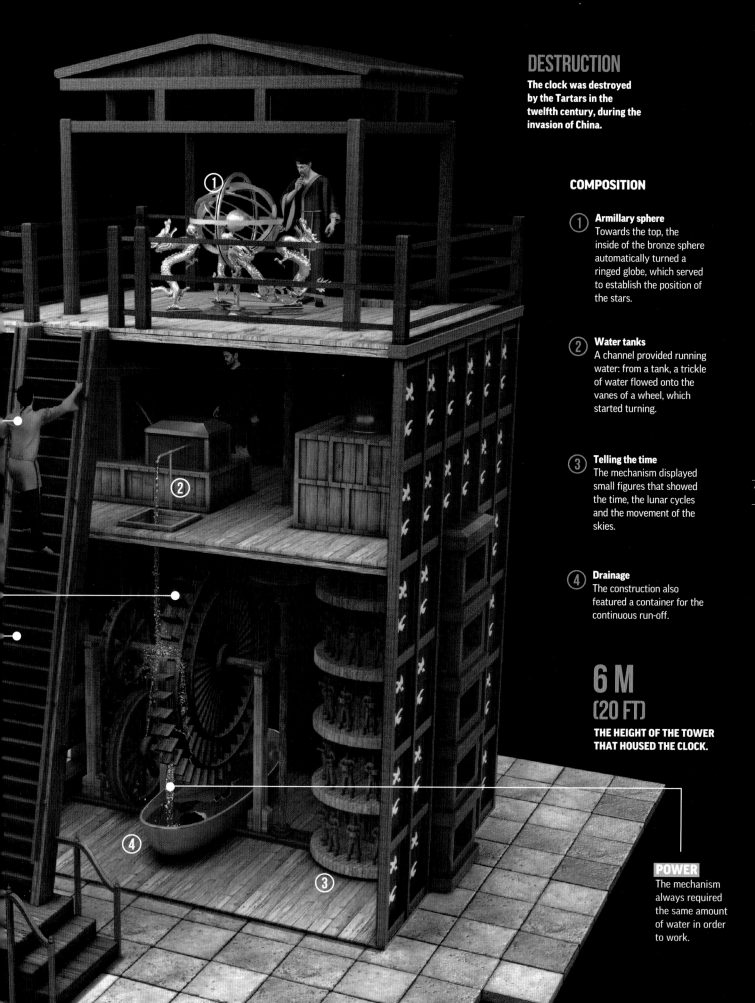

## DESTRUCTION

**The clock was destroyed by the Tartars in the twelfth century, during the invasion of China.**

## COMPOSITION

**① Armillary sphere**
Towards the top, the inside of the bronze sphere automatically turned a ringed globe, which served to establish the position of the stars.

**② Water tanks**
A channel provided running water: from a tank, a trickle of water flowed onto the vanes of a wheel, which started turning.

**③ Telling the time**
The mechanism displayed small figures that showed the time, the lunar cycles and the movement of the skies.

**④ Drainage**
The construction also featured a container for the continuous run-off.

# 6 M
## (20 FT)

**THE HEIGHT OF THE TOWER THAT HOUSED THE CLOCK.**

**POWER**
The mechanism always required the same amount of water in order to work.

# COPERNICUS AND GALILEO

In just 150 years, our view of the Universe was changed dramatically. Towards the end of the fifteenth century, Nicolaus Copernicus turned astronomy on its head with the philosophical assertion that Earth was not the center of the Universe, as upheld by the Church. Despite being outlawed, astronomers such as Tycho Brahe, Johannes Kepler and Galileo Galilei perfected Copernicus' theory.

## THE COPERNICAN REVOLUTION

Prior to the publication of Nicolaus Copernicus' theories in 1543, the geocentric theory of the Universe, developed by Ptolemy in the second century, was the commonly accepted belief. Copernicus, who discovered anomalies in Ptolemy's system, mathematically reached the conclusion that Earth moved, turning on its own axis in addition to orbiting the Sun (heliocentrism). He also revealed that the Earth's axis was tilted. His ideas gave rise to a new period: they marked the beginning of the scientific revolution and served as the basis for modern astronomy.

### TELESCOPE
This instrument was developed a year before in Holland, but Galileo was the first to use it to view the sky, in 1609. At the time, it was known as a spyglass.

### SPHERICAL ASTROLABE
Also known as an armillary sphere, it fixed the position of the stars in space.

### THE LEGACY OF COPERNICUS
Many of Copernicus' followers worked in the royal courts. Among them Tycho Brahe from Denmark, Johannes Kepler from Germany and Galileo Galilei from Italy.

### DE REVOLUTIONIBUS ORBIUM COELESTIUM
Copernicus never considered publishing his work, *On the Revolutions of the Heavenly Spheres*, but his pupil, Rheticus, was convinced. It was a decision that would change history.

### GALILEO GALILEI
The astronomer was forced to retract his ideas, but just afterwards stated: 'And yet it moves.'

**THE ECCLESIASTICAL POWERS DID
NOT ACCEPT GALILEO'S IDEAS AND HE
WAS CONVICTED OF HERESY IN 1633.**

## WHAT GALILEO SAW

Florentine astronomer and physicist Galileo Galilei
(1564–1642) was one of the fiercest defenders of
heliocentrism. He tried to convince ecclesiastical
sceptics that there were mountains on the Moon
and that Jupiter had several of its own moons.

### DOGE OF VENICE

When Galileo presented
his telescope to the Doge
of Venice, the latter
became interested in its
military uses.

### SENATORS

Galileo gave the senators
of Venice the rights
to manufacture the
telescope (although
strictly speaking, it was
not his invention).

### SCIENTIFIC METHOD

Galileo was one of the
creators of modern scientific
thinking, when he combined
inductive reasoning and
mathematical deduction.
Ever since this has been the
method used in physics.

# NEWTON AND THE UNIVERSAL GRAVITATION

**Famed for his theory of universal gravitation, Englishman Isaac Newton (1643–1727) is considered one of the greatest scientists of all times.** His field of research and his contributions to science exceed the boundaries of astronomy. Physicist, philosopher, inventor and mathematician, he is also famed for his works on light and optics, and the development of mathematical calculus.

## A REVOLUTIONARY THEORY

In his work, *Philosophiae naturalis principia mathematica*, published in 1687, Newton described one of the theories that has most influenced modern astronomy: the law of universal gravitation. This asserts that bodies exert a mutual force of attraction, depending on their mass and the distance between them. Furthermore, he suggested that all bodies are governed by the same physical laws of the Universe. Using his law of gravitation, Newton explained the movement traced by the planets in the Solar System, and demonstrated that the gravitation pull of the Sun and the Moon on the Earth's oceans were responsible for the tides.

### NEWTON AND THE APPLE

Legend has it that Newton started to develop his theory of universal gravitation when studying how an apple fell from the tree, asking himself why objects fall to the ground.

## THE LAWS OF MOTION

Newton put forward a series of laws of motion that were named after him, which explain the movement of bodies, their effects and their causes. These principles are: the law of inertia; the law of interaction and force; and the law of action-reaction.

## 11.2 KM/S
### (6.92 MI/S)

**IS THE SPEED ANY OBJECT REQUIRES TO ESCAPE EARTH'S GRAVITY... AND NEVER RETURN.**

## JOHANNES KEPLER (1571–1630)

This German astronomer laid the foundation for Newton's works. He established that the orbits of the planets around the Sun are shaped elliptically. He also discovered that the closer a planet is to the Sun, the faster it moves. He created the laws of planetary movement that made it possible to join together and forecast the movement of the stars.

### KEPLER'S THREE LAWS

1. The planets move elliptically around the Sun.

2. The areas cleared by the planets' radii are proportionate to the time it takes them to travel the perimeter of these areas.

3. The square of the orbital period of a planet is proportional to the cube of its average distance from the Sun.

## TELESCOPE

As part of his studies on the nature of light, Newton created a kind of reflecting telescope.

## 9.8 M/S²
### (32 FT/S²)

**IS THE ACCELERATION GENERATED BY THE FORCE OF GRAVITY ON EARTH, THAT IS, THE SPEED WITH WHICH BODIES ARE ATTRACTED TO THE EARTH'S SURFACE.**

## SPECTRAL LIGHT

Around 1666, Newton was able to demonstrate that light, considered white, was actually made up of colored light whose paths diverted at different angles (due to refraction) when passing through a glass prism. In 1671, Newton baptized the image created as the 'spectrum of light.'

# THE IMPACT OF RELATIVITY

Over the course of the first decades of the twentieth century, physicist Albert Einstein shook the world with his (special and general) theory of relativity, implying significant changes to the concepts of space, time and gravity that had been considered as well established since the times of Newton and Galileo. Similarly, our perception of the origin, evolution and structure of the Universe was to change just as dramatically.

## THE THEORY OF RELATIVITY

In 1905, Albert Einstein proposed the special theory of relativity that time and space are not absolute and independent of one another, but that they 'merge' in a constant four-dimensional dynamic known as 'space-time,' that can bend itself as if it were an elastic band. A decade later, in 1916, Einstein incorporated gravity into his theory (general relativity) and concluded that gravity is the result of the space-time curvature. Einstein's ideas would go on to suggest the concept of the Big Bang theory, in addition to the existence of black holes.

**SOLAR MASS**
At the heart of the scene, the Sun curves the surrounding space-time and 'diverts' the light of the stars, which are seen in an inaccurate position.

**LUNAR SHADOW**
When positioned between the Sun and Earth, the shadow of the Moon blocks out solar light and the resulting eclipse allows stars to be seen during the day.

**CURVED LINE**
The path of stellar light is curved or 'deflected' by the Sun.

① **EXPEDITIONS**
In 1919, two scientific missions used a solar eclipse to test Einstein's theory: if it were correct, the light of the stars should move at a certain angle when passing close to the Sun.

**1919 ECLIPSE**
Just a few minutes were enough. On 29 May 1919, two British expeditions confirmed Einstein's theoretical prediction. One viewed the total eclipse of the Sun from northern Brazil, and another from western Africa.

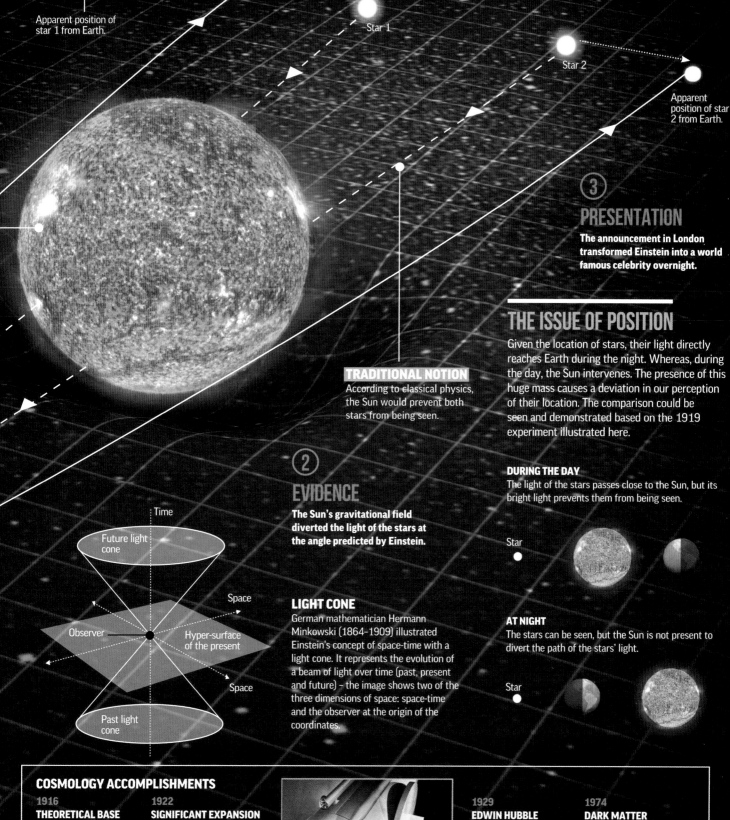

Apparent position of star 1 from Earth.

Star 1

Star 2

Apparent position of star 2 from Earth.

③

## PRESENTATION

**The announcement in London transformed Einstein into a world famous celebrity overnight.**

---

## THE ISSUE OF POSITION

Given the location of stars, their light directly reaches Earth during the night. Whereas, during the day, the Sun intervenes. The presence of this huge mass causes a deviation in our perception of their location. The comparison could be seen and demonstrated based on the 1919 experiment illustrated here.

### DURING THE DAY
The light of the stars passes close to the Sun, but its bright light prevents them from being seen.

Star

### AT NIGHT
The stars can be seen, but the Sun is not present to divert the path of the stars' light.

Star

### TRADITIONAL NOTION
According to classical physics, the Sun would prevent both stars from being seen.

②

## EVIDENCE

**The Sun's gravitational field diverted the light of the stars at the angle predicted by Einstein.**

### LIGHT CONE
German mathematician Hermann Minkowski (1864–1909) illustrated Einstein's concept of space-time with a light cone. It represents the evolution of a beam of light over time (past, present and future) – the image shows two of the three dimensions of space: space-time and the observer at the origin of the coordinates.

Time

Future light cone

Space

Observer

Hyper-surface of the present

Space

Past light cone

---

## COSMOLOGY ACCOMPLISHMENTS

**1916**
### THEORETICAL BASE
Albert Einstein's general theory of relativity became the most precise theoretical framework to describe

**1922**
### SIGNIFICANT EXPANSION
Russian mathematician Alexander Friedmann suggested that the Universe expands: laying the groundwork for the

**1929**
### EDWIN HUBBLE
Proposed that the Universe comprised numerous separate galaxies and that the Milky Way is just one

**1974**
### DARK MATTER
Soviet scientists demonstrate that the majority of the Universe is made up of dark matter, a fact that had been

# LARGE HADRON COLLIDER

**The Large Hadron Collider (LHC) is the largest scientific instrument ever made by mankind,** and offers a great insight into understanding how the Universe works. Its purpose is to smash high-speed, energized particles into one another, in order to obtain data about the basic forces of the Universe and to discover new elementary particles.

## THE LHC COMPLEX

Located on the border between Switzerland and France, the LHC consists of several elements: several ring-shaped tunnels measuring almost 9 km (5.6 mi) in diameter, which raise the energy of particles to levels never before induced artificially, and superconducting magnets, to direct and shoot them. Six experiments have been carried out to analyze the results of the collisions.

## CMS DETECTOR

This instrument weighs around 12,500 tons and is designed to analyze, during collisions between highly energized protons, the particles generated (such as photons, muons and other elementary particles) and aspects such as their mass, energy and speed.

### SUPERCONDUCTING MAGNETS

Cooled to almost absolute zero (-273° C/ -459° F) with liquid nitrogen, the magnets are the largest that have ever been built. They impart high energy to the particles and guide them.

### MUON IDENTIFIER

Permits the detection of this fundamental particle and allows for the measurement of its mass and velocity.

**(1) Particle accelerator**
A linear particle accelerator separates atom nuclei from their electrons to form ions. Some ions contain just one proton (hydrogen ions), but others have more than one (such as lead ions). These ions are directed to the underground complex.

**(2) Speed**
The ions are accelerated to reach speeds close to that of light.

**(3) Raising the energy**
Powerful impulses of radio waves raise the energy of the ions to 400 billion electron volts.

**(4) Collision**
Streams of billions of now very highly energized ions are introduced into the LHC accelerator, some in one direction and others in the opposite direction. Superconducting magnets then increase their energy tenfold before particles are made to collide with each other.

### INPUT
Of particles that will collide.

### HADRON CALORIMETER
Records the energy of the hadrons and analyzes their interactions with atomic nuclei.

### ELECTROMAGNETIC CALORIMETER
Precisely measures the

21.5 m (71 ft)

15 m (49 ft)

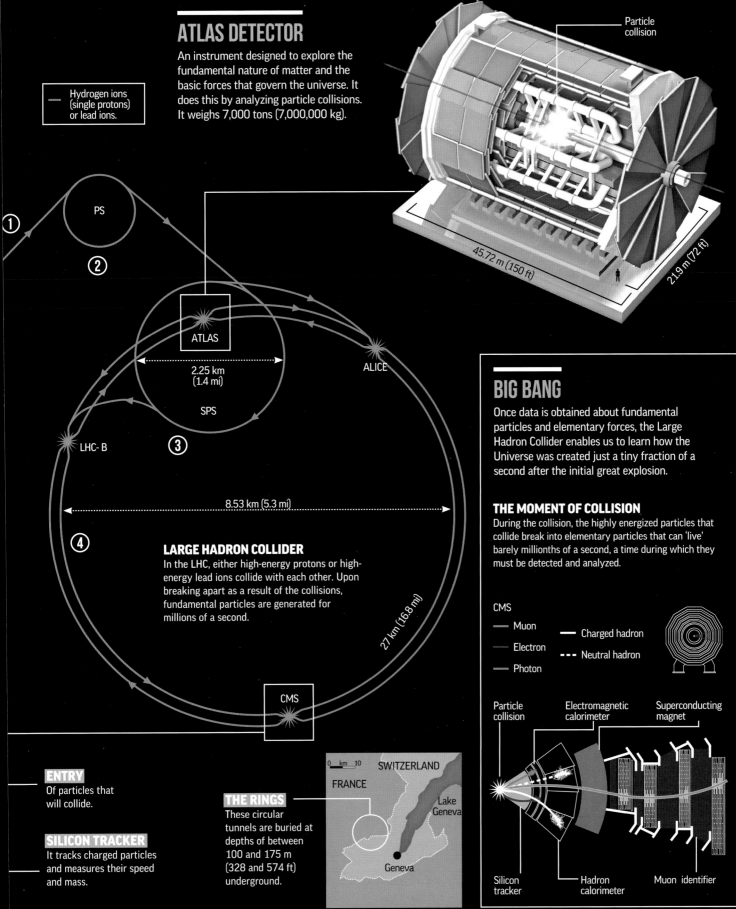

# ATLAS DETECTOR

An instrument designed to explore the fundamental nature of matter and the basic forces that govern the universe. It does this by analyzing particle collisions. It weighs 7,000 tons (7,000,000 kg).

Hydrogen ions (single protons) or lead ions.

Particle collision

45.72 m (150 ft)

21.9 m (72 ft)

① ② PS

ATLAS

2.25 km (1.4 mi)

SPS

ALICE

LHC- B

③

8.53 km (5.3 mi)

④

27 km (16.8 mi)

## LARGE HADRON COLLIDER

In the LHC, either high-energy protons or high-energy lead ions collide with each other. Upon breaking apart as a result of the collisions, fundamental particles are generated for millions of a second.

CMS

# BIG BANG

Once data is obtained about fundamental particles and elementary forces, the Large Hadron Collider enables us to learn how the Universe was created just a tiny fraction of a second after the initial great explosion.

## THE MOMENT OF COLLISION

During the collision, the highly energized particles that collide break into elementary particles that can 'live' barely millionths of a second, a time during which they must be detected and analyzed.

CMS

— Muon
--- Electron
— Photon
— Charged hadron
--- Neutral hadron

Particle collision

Electromagnetic calorimeter

Superconducting magnet

Silicon tracker

Hadron calorimeter

Muon identifier

**ENTRY**
Of particles that will collide.

**SILICON TRACKER**
It tracks charged particles and measures their speed and mass.

0   km   10

SWITZERLAND

FRANCE

Lake Geneva

Geneva

**THE RINGS**
These circular tunnels are buried at depths of between 100 and 175 m (328 and 574 ft) underground.

# SPACE OBSERVATORY

The earliest astronomical observatories date back to the time of ancient Babylon, but it barely glimpsed the stars. In the last century, the evolution of the technology has allowed us to build powerful telescopes in zones of low light pollution with the capacity to detect stars and planets to billions of kilometers.

**DOME**
Protects and perceives any climate change through thermal sensors.

## THE PARANAL OBSERVATORY

The Very Large Telescope (VLT) is one of the most advanced astronomical observatories in the world. It has four telescopes that make it possible, for example, to see the flame of a candle on the surface of the moon. It is operated by a scientific consortium composed of eight countries in Europe and one of its objectives is to find new worlds around other stars.

### WEATHER CONDITIONS

The Cerro Paranal (Region II of Chile) is located in the driest part of the Atacama Desert, where conditions for astronomical observation are extraordinary. It is a mountain 2,635 m (8,645 ft) high and offers nearly 350 cloudless nights per year with an unusual atmospheric stability.

**SECONDARY MIRROR**
1.2 m (4 ft) in diameter.

### THE TELESCOPE

The main feature of the VLT is its revolutionary optical design. Thanks to the active and adaptive optics, you get a resolution similar to being in space.

## 20,000 M²
### (215,278 SQ FT)

**TOTAL SURFACE.**

## OTHER OBSERVATORIES

**VERY LARGE ARRAY**
This radioastronomical observatory is in New Mexico and has 27 independent antennas.

**LA SILLA**
In the desert of Atacama, Chile. Its powerful telescopes are specialized in the search for extrasolar planets.

**GEMINI**
It consists of two twin telescopes located in both hemispheres of the planet (Hawaii and Chile) to cover the entire terrestrial sky.

## TELESCOPIC UNITS

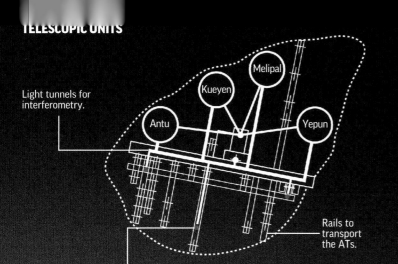

Light tunnels for interferometry.

**Antu** **Kueyen** **Melipal** **Yepun**

Rails to transport the ATs.

**Auxiliary telescopic units (ATs).** There are three of these, 1.8 m (6 ft) in diameter each, and they are used for interferometry.

## THE COMPLEX

Finished in 2006, the very large telescope (VLT) has four reflecting telescopes 8.2 m (27 ft) in diameter that can observe objects 4,000 million times weaker than can normally be seen with the naked eye. It also has three movable auxiliary telescopes of 1.8 m (6 ft) in diameter which, when combined with the large telescopes, produce what is called interferometry: a simulation of the power of a mirror 16 m (52 ft) in diameter and the resolution of a telescope of 200 m (656 ft), which can distinguish an astronaut on the Moon.

## OPTICAL ADAPTATION

To counteract the blurring effects of the Earth's atmosphere, the VLT has an active optical system featuring 150 pistons that move mirror segments to realign light into sharp images.

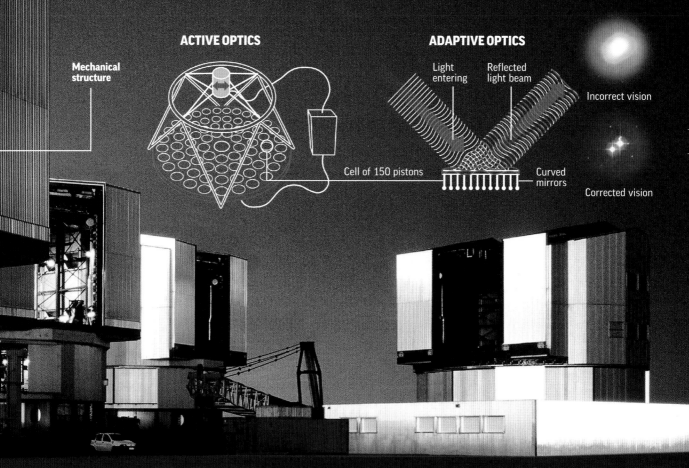

**Mechanical structure**

### ACTIVE OPTICS

### ADAPTIVE OPTICS

Light entering

Reflected light beam

Incorrect vision

Cell of 150 pistons

Curved mirrors

Corrected vision

### MAUNA KEA
A resort built on an inactive volcano, in Hawaii, with UK, French-American and American observatories.

### ALMA
The Atacama Large Millimeter / submillimeter Array has 66 high-precision antennas to observe millimeter

### CANARIAS
This Spanish archipelago has two important observatories: Teide and Roque de los Muchachos.

# CELESTIAL CARTOGRAPHY

**As on the Earth, so in heaven. Just as terrestrial maps help us find locations on the surface of the planet,** star charts use a similar coordinate system to indicate various celestial bodies and locations. Planispheres, or star wheels, are based on the idea of a celestial sphere (an imaginary globe on which the stars appear to lie and that surrounds the planet). Two common types are polar and bimonthly star maps.

## THE CELESTIAL SPHERE

The celestial sphere is imagined to extend around the Earth and forms the basis for modern star cartography. The sphere is divided into a network of lines and coordinates corresponding to those used on the Earth, allowing an observer to locate constellations on the sphere. The celestial equator is a projection of the Earth's equator, the north and south celestial poles align with the axis of the Earth, and the elliptic coincides with the path along which the Sun appears to move.

**MAP OF THE STELLAR NORTH POLE**

## MEASURING DISTANCES

Once a star or constellation has been located in the sky, hands and arms can serve as simple measuring tools. A single extended finger, shown in the first illustration, can form a one-degree angle from the observer's line of sight and is useful for measuring short distances between stars. The closed palm of the hand forms a 10° angle, and the open hand measures 20°.

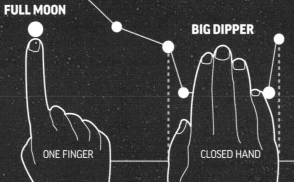

**FULL MOON**

**BIG DIPPER**

**THE SQUARE OF PEGASUS**

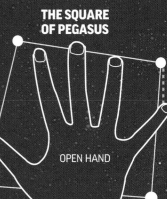

ONE FINGER

CLOSED HAND

OPEN HAND

## HOW TO READ A MAP OF THE SKY

Astronomers divide the celestial sphere into sections, allowing them to study the sky in detailed, systematic ways. These maps can show a particular region observed from a certain place at a certain time, or they can merely concentrate on a specific location. To specify the position of a point on the surface of the Earth, the geographic coordinates called latitude and longitude are used. With the celestial sphere, declination and right ascension are used instead. Observers located at the equator see the celestial equator pass directly over their heads.

Star magnitudes

Constellations

Milky Way

## STELLAR MOVEMENTS

The visible regions of the celestial sphere and the ways in which stars move through the sky depend upon the observer's latitude. As an observer moves north or south, the visible portion of the celestial sphere will change. The elevation of the north or south celestial pole above the horizon determines the apparent motion of the stars in the sky.

**AT THE POLES**
At the North Pole, the stars appear to rotate around the observer's head. The effect is the same at the South Pole but in the opposite direction.

**IN MIDDLE LATITUDES**
some stars can be seen all year long, but others are only visible during certain months.

**AT THE EQUATOR**
stars can be seen throughout the year, rising in the east and setting in the west.

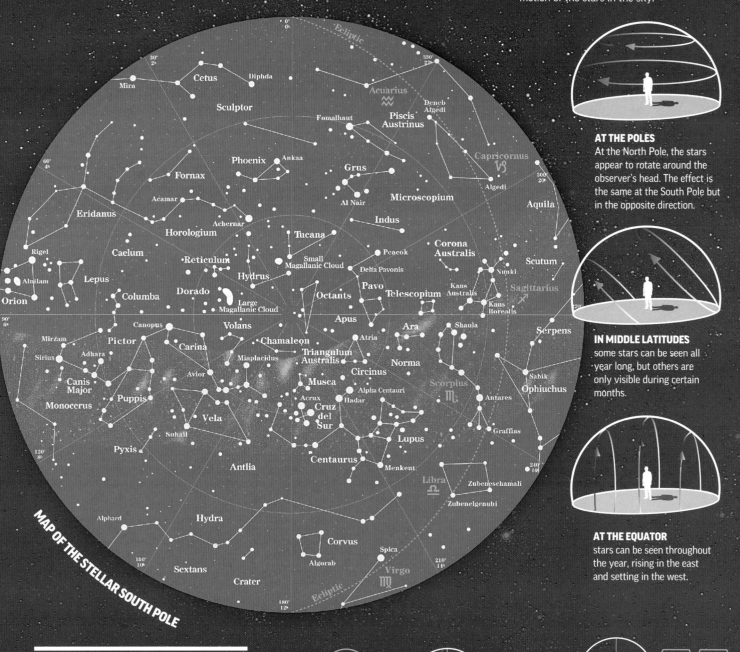

MAP OF THE STELLAR SOUTH POLE

Mira · Cetus · Diphda · Sculptor · Acuarius · Fomalhaut · Piscis Austrinus · Deneb Algedi · Capricornus · Phoenix · Ankaa · Grus · Fornax · Microscopium · Algedi · Acamar · Al Nair · Aquila · Eridanus · Achernar · Indus · Horologium · Tucana · Caelum · Reticulum · Small Magallanic Cloud · Peacok · Corona Australis · Scutum · Hydrus · Delta Pavonis · Pavo · Telescopium · Numki · Rigel · Dorado · Octants · Apus · Kaus Australis · Kaus Borealis · Sagittarius · Almilam · Large Magallanic Cloud · Volans · Ara · Shaula · Serpens · Orion · Lepus · Columba · Chamaleon · Atria · Norma · Sabik · Canopus · Triangulum Australis · Circinus · Scorpius · Ophiuchus · Mirzam · Pictor · Carina · Miaplacidus · Musca · Alpha Centauri · Antares · Sirius · Adhara · Avior · Acrux · Hadar · Canis Major · Puppis · Cruz del Sur · Lupus · Graffias · Monocerus · Vela · Suhail · Centaurus · Menkent · Pyxis · Antlia · Zubeneschamali · Zubenelgenubi · Libra · Alphard · Hydra · Corvus · Spica · Sextans · Algorab · Virgo · Crater · Ecliptic

## DIFFERENT TYPES OF CHARTS

Throughout the year, different constellations are visible because the Earth moves along its orbit. As the Earth's place in its orbit changes, the night side of the planet faces different regions of space. To compensate for this shifting perspective, there are various kinds of planispheres: north and south polar maps and bimonthly equatorial maps.

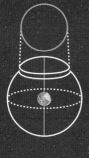

**POLAR**
The celestial sphere is generally divided into two polar maps: north and south.

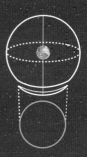

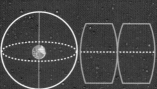

**EQUATORIAL**
Six bimonthly maps depict all 88 constellations, which can be seen over the course of the year.

# FROM THE BACK YARD

Today, thanks to powerful domestic binoculars and telescopes, space can be seen at an acceptable quality and, with the help of a stellar map, it is possible to recognize galaxies, nebulae, star clusters, planets and much more. It is important to become familiarized with the night sky first to make the most of observation time.

## BASICS

Before heading off to view the skies, it is essential to ensure you have everything you need. The basic equipment for any budding observer includes binoculars, or even better, a telescope, star maps (planisphere) and a notebook. A compass is also handy for establishing the four directional points, in addition to a lamp.

Planisphere

Compass

Lamp with red cellophane to preserve dark-adapted vision

BARREL

OPTICS TUBE

## HOW TO SEE THE MOON

Observers can use instruments such as binoculars and telescopes to enjoy different views of astronomical objects. The Moon is a particularly good target for beginners to study because it is generally very bright.

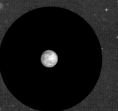

**Moon**
Normal view.

**10 times larger**
View with binoculars.

**50 to 100 times larger**
View with a telescope.

FOCUS WHEEL

FOCUS EYEPIECE

ADJUSTMENT NUT

TRIPOD ADAPTOR

## MOVING CONSTELLATIONS

Just as the Sun rises in the east in the morning and sets in the evening, the stars and planets do exactly the same. However, it is not really the stars and planets moving but actually the Earth spinning that creates this cosmic optical illusion.

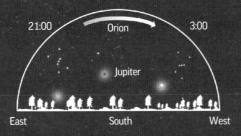

21:00   Orion   3:00

Jupiter

East    South    West

## VIEWABLE OBJECTS

In addition to stars and planets, communication and spy satellites, planes, comets and meteorites can all be seen in the sky. They are recognizable due to both their shape and their movement.

### SHOOTING STARS
Very short bursts of light that last no longer than a fraction of a second.

### MOON
The illuminated side of the Moon faces Earth and is always visible, at least partially, apart from during a new Moon.

### VENUS
Generally, it can be seen on the horizon at dusk.

### COMMUNICATION SATELLITES
The largest shine brighter than some stars and others take a long time to cross the night sky.

### COMETS
Comets visible to the naked eye can be seen every couple of years, and can be seen for weeks or months at a time.

LENS

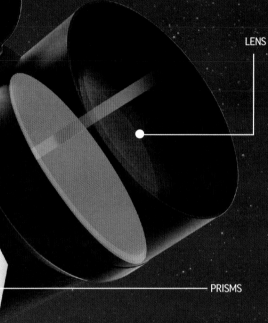

PRISMS

## JOINING THE DOTS

A constellation is a group of stars that, when seen from a certain angle, appear to take on a given shape and appear close together. However, in reality, they are separated by huge distances.

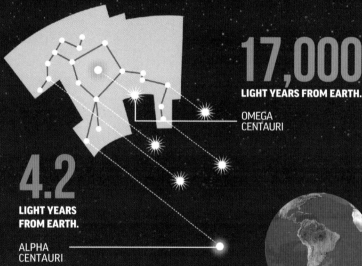

# 17,000
**LIGHT YEARS FROM EARTH.**

OMEGA CENTAURI

# 4.2
**LIGHT YEARS FROM EARTH.**

ALPHA CENTAURI

## MEASUREMENT METHODS

A celestial planisphere is a circular map of the constellations used to locate stars. To identify one in particular, your arms and your body can be used to measure the direction and height of an object against the horizon.

**MEASURING DIRECTION**

**MEASURING HEIGHT**

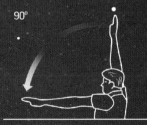
90°

45°

HORIZON

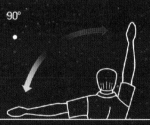
90°

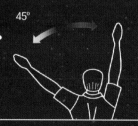
45°

Planispheres indicate the cardinal direction of a star. Position your arms at 90° using north or south as a starting point.

A star to the southeast may be located at 45°. Combine directional angles with height measurements using your arms.

Using the horizon as a starting point, extend one of your arms on this line and the other perpendicularly.

The easiest way to measure a 45° angle is to move your arm from the horizon to halfway through the 90° angle.

**SPACE SHUTTLE ENDEAVOUR**
Launch of Endeavour on September 30,
1994 with a crew of six NASA astronauts
and the Space Radar Laboratory-2 (SRL-2).

## CHAPTER 5

# THE SPACE RACE

The launch of the small Soviet satellite Sputnik in 1957 represented the beginning of the space race between the two great world powers of the time: the United States and the USSR. The Apollo program, with the arrival of man to the Moon in 1969, was the first great milestone of space exploration.

# FROM FICTION TO REALITY

**Astronautics came about in the late nineteenth century, when the Russian Konstantin Tsiolkovsky (1857–1935) foresaw the ability of a rocket to overcome the force of gravity.** But the space race, the rivalry between world powers to conquer space, officially began in 1957 with the launch of the first Soviet artificial satellite, Sputnik I.

## SPUTNIK I

It consisted of an aluminium sphere of 58 cm (2 ft) in diameter, and for 21 days sent information on cosmic radiation, meteorites and the density and temperature of the Earth's upper atmosphere. It was destroyed by aerodynamic drag upon entering the atmosphere 57 days later.

**83.6 KG**
**(184 LB)**

**WEIGHT ON THE GROUND**

0.58 m
(2 ft 7 in)

### THE FIRST
In 1917 in Germany, Romanian Hermann Oberth (1894–1989) projected a liquid-fuelled rocket, which promoted the idea of spaceflight.

### THE SECOND
American Robert Goddard (1882–1945) designed a rocket 3 m (10 ft) high. On ignition, it rose to 12 m (39 ft) and then crashed 56 m (184 ft) away.

### THE THIRD
German physicist Wernher von Braun (1912–77), created the Saturn V rocket for NASA: the rocket that took men to the Moon in 1969 and 1972.

### TECHNICAL DATA SHEET

**Launch**
October 1957

**Orbital altitude**
600 km (373 mi)

**Orbital period**
97 minutes

**Weight**
83.6 kg (184 lb)

**Country**
USSR

**ANTENNAS**
Sputnik I had four antennae of between 2.4 and 2.9 m (8 to 9.5 ft) in length.

**1609**
**GALILEO**
Built the first astronomical telescope and observed lunar craters.

**1798**
**CAVENDISH**
Shows that the law of gravity is true for any pair of bodies.

**1806**
**ROCKETS**
The first were invented for military use. They were used in 1814 in an air strike.

**1838**
**DISTANCE**
The distance to the star 61 Cygni, taking Earth's orbit as a reference, was measured.

**1926**
**FIRST ROCKET**
Robert Goddard launches the first liquid-fuelled rocket.

the dog Laika. The dog was connected to a machine that recorded its vital signs and an air-regeneration system provided it with oxygen.

## TECHNICAL DATA SHEET

**Launch**
Nov 1957

**Orbital altitude**
1,660 km (1,031 mi)

**Orbital period**
103.7 minutes

**Weight**
508 kg (1,120 lb)

**Country**
USSR

# 508 KG
## (1,120 LB)
**WEIGHT ON THE GROUND**

4 m (13 ft)

2 m (6 ft 6 in)

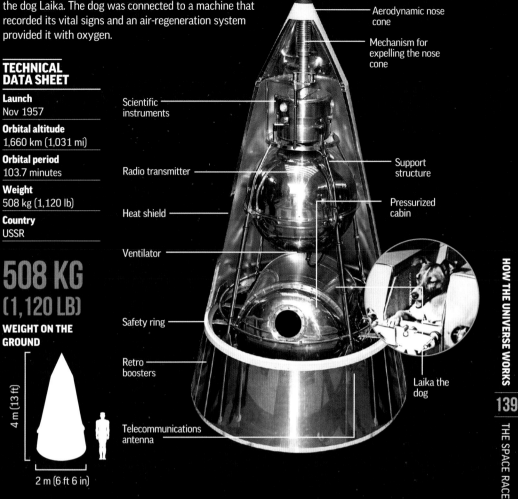

Aerodynamic nose cone

Mechanism for expelling the nose cone

Scientific instruments

Support structure

Radio transmitter

Pressurized cabin

Heat shield

Ventilator

Safety ring

Retro boosters

Laika the dog

Telecommunications antenna

# 14 KG
## (31 LB)
**WEIGHT ON THE GROUND**

0.8 m (2 ft)

# EXPLORER I

United States developed its first satellite, Explorer I, in 1958, launched from Cape Canaveral. It was a small cylindrical vessel 15 cm (6 in) in diameter that measured cosmic radiation and meteorites for 112 days, allowing the discovery of the Van Allen Belts.

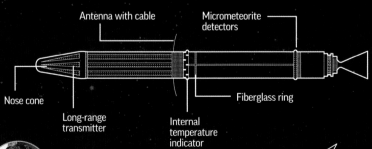

Antenna with cable

Micrometeorite detectors

Nose cone

Fiberglass ring

Long-range transmitter

Internal temperature indicator

## TECHNICAL DATA SHEET

**Launch**
Feb. 1958

**Orbital altitude**
2,550 km (1,585 mi)

**Orbital period**
114.8 minutes

**Weight**
14 kg (31 lb)

**Organization**
NASA

**1927**
**SPACE SOCIETY**
The Society for Space Travel was founded on July 5th in Germany.

**1932**
**VON BRAUN**
He began his research on rockets for the German army.

**1947**
**ROCKET PLANE**
Chuck Yeager breaks the sound barrier aboard the X-1 rocket plane.

**1949**
**BUMPER**
First two-stage rocket, which reached an altitude of 393 km (244 mi).

**1957**
**SPUTNIK I**
On October 4th, the Soviet Union launched the first satellite into space.

# NASA

The National Astronautics and Space Administration (NASA) is the US space agency. It was created in 1958 as part of the "space race" being contested with the then Soviet Union, and it scheduled all national activities related to space exploration. NASA has facilities throughout the country, including at the Kennedy Space Center.

## NASA BASES

NASA has facilities throughout the United States that develop research, flight simulation and astronaut training. NASA headquarters are in Washington DC, and the Flight Control Center in Houston, Texas, is one of the locations where the Deep Space Network operates. It is a communications system with three locations in the world: Houston, Madrid and Canberra, which can capture signals in all directions and covers 100 percent of the Earth's surface.

### AMES RESEARCH CENTER
Founded in 1939, it is the experimental base for many missions. It is equipped with simulators and advanced technology.

### LYNDON B. JOHNSON CONTROL CENTER
The center at Houston selects and trains astronauts and controls the take-off and landing of flights.

### MARSHALL SPACE FLIGHT CENTER
Handles equipment transport, propulsion systems and space-shuttle launch operations.

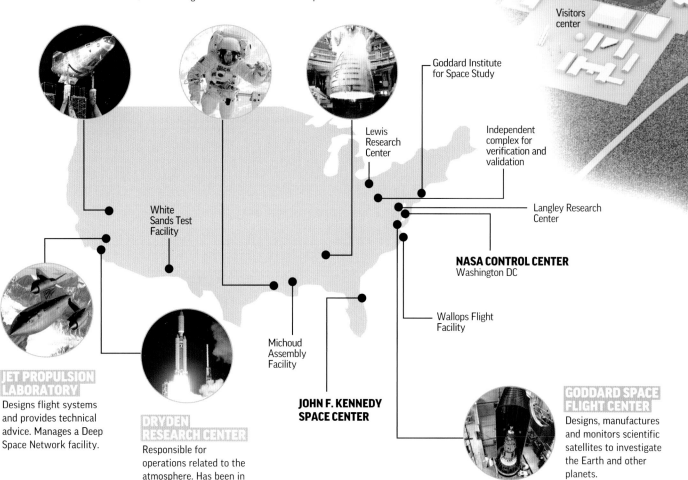

INDIAN RIVER

**SHUTTLE LANDING FACILITY**

Visitors center

Goddard Institute for Space Study

Lewis Research Center

Independent complex for verification and validation

Langley Research Center

White Sands Test Facility

**NASA CONTROL CENTER**
Washington DC

Wallops Flight Facility

Michoud Assembly Facility

**JOHN F. KENNEDY SPACE CENTER**

### JET PROPULSION LABORATORY
Designs flight systems and provides technical advice. Manages a Deep Space Network facility.

### DRYDEN RESEARCH CENTER
Responsible for operations related to the atmosphere. Has been in operation since 1947.

### GODDARD SPACE FLIGHT CENTER
Designs, manufactures and monitors scientific satellites to investigate the Earth and other planets.

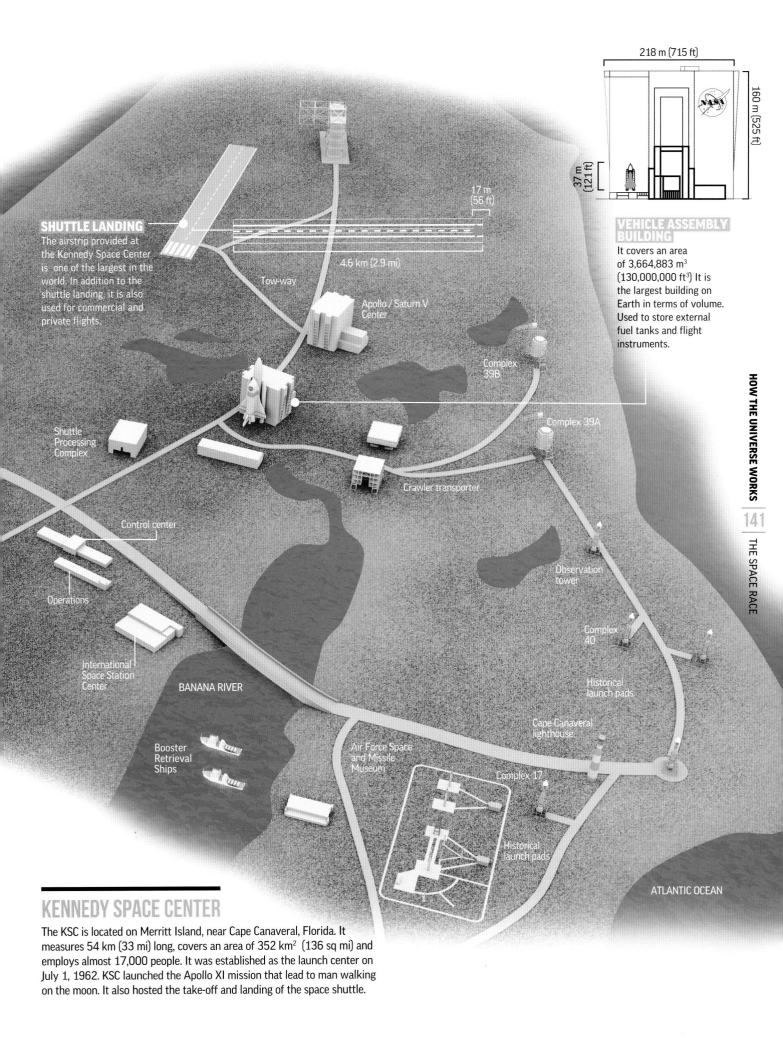

**SHUTTLE LANDING**
The airstrip provided at the Kennedy Space Center is one of the largest in the world. In addition to the shuttle landing, it is also used for commercial and private flights.

17 m (56 ft)

4.6 km (2.9 mi)

Tow-way

Apollo / Saturn V Center

Complex 39B

Complex 39A

Shuttle Processing Complex

Crawler transporter

Control center

Operations

Observation tower

Complex 40

Historical launch pads

International Space Station Center

BANANA RIVER

Cape Canaveral lighthouse

Booster Retrieval Ships

Air Force Space and Missile Museum

Complex 17

Historical launch pads

ATLANTIC OCEAN

218 m (715 ft)

160 m (525 ft)

37 m (121 ft)

**VEHICLE ASSEMBLY BUILDING**
It covers an area of 3,664,883 m³ (130,000,000 ft³) It is the largest building on Earth in terms of volume. Used to store external fuel tanks and flight instruments.

# KENNEDY SPACE CENTER

The KSC is located on Merritt Island, near Cape Canaveral, Florida. It measures 54 km (33 mi) long, covers an area of 352 km² (136 sq mi) and employs almost 17,000 people. It was established as the launch center on July 1, 1962. KSC launched the Apollo XI mission that lead to man walking on the moon. It also hosted the take-off and landing of the space shuttle.

# CONTROL FROM EARTH

**Monitoring of astronauts' activity is done from operations centers. In the United States, NASA coordinates manned missions from the Mission Control Center located in the Johnson Space Center (JSC)in Houston, Texas. The unmanned missions are supervised from the Jet Propulsion Laboratory in Los Angeles, California.**

## BUILDING 30

In this building of the JSC are the different rooms of the Mission Control Center, from which all NASA manned space flight is monitored. Remodeled in 2006, Flight Control Room 1 (FCR-1) has been the control center of both the Gemini and Apollo space programs and the Space Shuttle. Today it is in charge of supervising the activity of the International Space Station (ISS).

### JSC, HOUSTON
The complex, also known as Space City, was built in 1963 on 1,620 acres, in the Clear Lake Area of Houston. It consists of one hundred buildings.

### CONSOLES
The classic consoles of the Operations Control Room form desks with an area for more than one monitor. Countertops and drawers provide work areas.

**Folding table**
For supporting objects and books.

**Monitor**
To display data from spacecraft and other systems.

### FLIGHT DIRECTOR
Responsible for the countdown before liftoff and the flight plans.

**Rear sliding drawer**
To keep information and papers.

**Protective covering**
Prevents damage to the console system.

### ROWS
NASA control rooms have always been arranged in several rows of consoles, specialized in various fields: contact with the crew, energy systems, extravehicular exits, etc.

## THE BIG SCREENS

Several enormous screens dominate the Mission Control Center. They provide information on the location and orbital trajectory of spacecraft in flight, as well as other data. The screens are of vital importance for the operators, because they allow for the rapid reading of information to take action efficiently and prevent accidents.

## 365 DAYS
**PER YEAR AND 24 HOURS PER DAY THAT SPACE CONTROLS ARE PERFORMED.**

### SCREEN 1
Shows the location and path of spacecraft in orbit.

### SCREEN 2
Records the location of satellites and other objects in orbit.

### SURGEON
A doctor checks the astronauts' vital signs during the flight and establishes treatment if necessary.

## WHITE, BLUE, AND RED ROOMS

In 1998 NASA expanded the Building 30 with one more wing that included three new independent rooms: the White, used for Space Shuttle operations until 2011; the Blue (image below), dedicated to ISS operations until 2006; and the Red, a training room for flight controllers. The White and Blue FCR are upgraded to MCC-21 (21st Century Control Center).

# OTHER SPACE AGENCIES

**Activity for exploration of the cosmos expanded in 1975, when the European Space Agency (ESA) was created.** This intergovernmental organization has the largest investment budget after NASA. The Mir station, launched by the Russian Federal Space Agency (RKA), was 15 years in orbit and was a vital milestone for life in space. Other, younger agencies are Canada's CSA and Japan's JAXA.

## EUROPE IN SPACE

The ESA was established as an organization in 1975, when the European Space Research Organization (ESRO) was merged with the European Launch Development Organization (ELDO). It has conducted missions of considerable importance, such as Venus Express, Mars Express and the Ulysses probe, the latter in conjunction with NASA.

### EUROPEAN SPACE AGENCY

| Founded | Annual budget |
|---|---|
| 1975 | 5.75 billion euros |

| Members | Employees |
|---|---|
| 22 | 2,200 |

# OVER 200
**ARIANE ROCKET LAUNCHES MADE BY THE ESA SO FAR.**

### EUROPEAN LAUNCH BASE
Latitude: 5° North, 500 km (311 mi) north of the equator.
Being so near the equator makes it easier to launch rockets into high orbits. The area is almost uninhabited and free of earthquakes.

### KOUROU, FRENCH GUIANA

| Area | First operation |
|---|---|
| 850 km² (328 sq mi) | 1968 (as a French base) |

| Total cost | Employees |
|---|---|
| 1.6 billion euros | 600 |

## THE ARIANE FAMILY

The development of the Ariane rocket has led the ESA to become market leader in launches. Ariane is chosen by Japanese, Canadian and American manufacturers. The Ariane 6 is currently under development and its first launch is scheduled for 2021-2022.

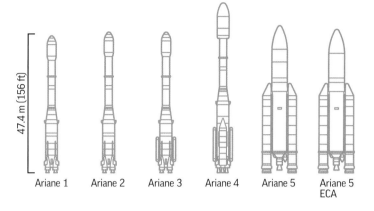

47.4 m (156 ft)

Ariane 1   Ariane 2   Ariane 3   Ariane 4   Ariane 5   Ariane 5 ECA

## CANADIAN SPACE AGENCY

The CSA was established in 1990, although Canada had already developed some astronautical activities before that. The first Canadian launch was in 1962 with the satellite Alouette I. The most important project is the Radarsat CSA, launched in November 1995. It provides information on the environment and is used in cartography, hydrology, oceanography and agriculture.

## RUSSIAN FEDERAL SPACE AGENCY

It was formed after the dissolution of the Soviet Union, and inherited the technology and launch sites. The new agency was responsible for the orbiting Mir Station, the forerunner to the International Space Station (ISS). The Mir was assembled in orbit by launching different modules separately, between 1986 and 1996. It was destroyed in a controlled manner on March 23rd, 2001.

POCKOCMOC

**Transportation route**

**ASSEMBLY BUILDING**
Once assembly is completed, the rocket is transferred to the platform.

**SPRINGBOARD**
After covering 3 km (1.9 mi) at 3.5 km/h (2.2 mph), the Ariane is ready for take-off.

**MIR STATION**
Mir housed both cosmonauts (Russia) and astronauts (US) in space.

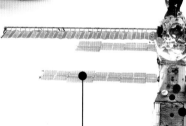

**PROGRESS-M**
Device for supplying food and fuel.

**TOWARDS THE FINAL DESIGN**
The rocket is directed to the integration building to finalize details.

**SOLAR PANELS**
They provide power for the station.

**MAIN MODULE**
For housing and overall control of the station.

## JAPANESE SPACE AGENCY

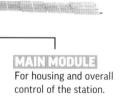

JAXA
Japan Aerospace Exploration Agency

On October 1st, 2003, three independent organizations were merged to form JAXA: the Institute of Space and Aeronautical Science (ISAS), the National Aerospace Laboratory (NAL) and the National Space Development Agency (NASDA). The highlight so far has been the Hayabusa mission, launched in May 2003. In November 2005, it became the first mission to land on an asteroid – the Itokawa. Despite the problems with the probe, the controllers managed to return it to Earth in 2010 with samples taken from the surface of the asteroid.

**SOYUZ ROCKET**
Belonging to the Russian agency, it is used to launch a spacecraft into orbit.

# RUSSIAN MISSIONS

**After early successes with small animals aboard satellites, the former USSR and the USA initiated the development of programs to launch humans into space.** The first astronaut to orbit the Earth was Yuri Gagarin in 1961, the only crew member of the Russian spaceship Vostok 1. Gagarin circled around the Earth in the capsule, launched into orbit by an SL-3 rocket, which allowed the cosmonaut to be ejected in an emergency.

## MEN IN SPACE

In Vostok I, the cosmonaut had virtually no control over the ship, which was operated remotely by Soviet engineers. The ship was made up of a spherical cockpit of 2.46 tons and 2.3 m (7 ft 6 in) in diameter. The one-man cockpit was mounted on the module containing the motor. Yuri Gagarin's re-entry was made using a parachute.

### VOSTOK 1

| **Launch** | **Weight** |
|---|---|
| April 1961 | 5,000 kg (11,023 lb) |
| **Orbital altitude** | **Organization** |
| 315 km (196 mi) | USSR |
| **Orbital period** | |
| 1 hour and 48 minutes | |

## 5,000 KG
## (11,023 LB)
**WEIGHT ON THE GROUND.**

4.5 m (15 ft)

Inflatable airlock

Nitrogen and oxygen tanks

Access gate

VHS antenna

Engine control

Retro boosters

### THE FIRST
On board Vostok 1, Yuri Gagarin (1934–68) was the first person to go into space, which made him a national hero. He died on a routine flight aboard a MiG -15 aircraft.

### THE FIRST WOMAN
The Russian Valentina Tereshkova (b. 1937) was the first woman cosmonaut. She travelled to space aboard the Vostok 6 in 1963. The mission made 48 orbits around the Earth in 71 hours of flight.

### SPACEWALK
Aleksei Leonov (b. 1934) was the first to perform a spacewalk, in March 1965. The ship that took him to outer space was the Voskhod 2. In 1975, he was made commander of the Apollo–Soyuz mission.

**1957**
**SPUTNIK 2**
Russian satellite launched 3 November, with the dog Laika.

**1958**
**EXPLORER 1**
First US Earth orbiter satellite.

**1958**
**NASA**
Foundation of the US space agency.

**1959**
**LUNA 1**
Launched by the USSR, reaches 6,000 km (3,728 mi) from the Moon.

**1959**
**LUNA 3**
Launched in October. Took pictures of the far side of the Moon.

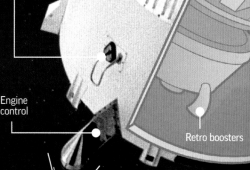
**1960**
**PUPPIES**
Strelka and Belka came back alive from a one day orbital voyage.

Nitrogen tanks

Cosmonaut

Cosmonaut ejection seat

## THE VOSTOK PROGRAMME

Vostok (Russian for 'east') was a Soviet space program that between April 1961 and June 1963 put six cosmonauts into orbit around the Earth. On June 16th, 1963, Vostok 6 took off carrying the world's first female cosmonaut, Valentina Tereshkova, on a joint flight with Vostok 5, piloted by Valery Bykovsky.

### VOSTOK MISSIONS PROGRAMME

| | |
|---|---|
| **Vostok 1** April 12th, 1961 | **Vostok 4** August 12th, 1962 |
| **Vostok 2** August 6th, 1961 | **Vostok 5** June 14th, 1963 |
| **Vostok 3** August 11th, 1962 | **Vostok 6** June 16th, 1963 |

### VOSTOK LAUNCH ROCKET

In order to leave Earth, the Vostok needed a launch rocket.

Crew module

First stage     Second stage     Third stage

② **The separation** is performed at 10:25 and the cosmonaut's re-entry begins at 10:35.

③ **The cosmonaut** ejects from the rocket by parachute.

① **The ship** takes off from Baikonur cosmodrome in Tyuratam, at 9:07.

## THE ROUTE

After take-off, it first crossed part of Siberia, then all of the Pacific, passed between Cape Horn and Antarctica and, after crossing the Atlantic, passed through the African sky over Congo. The capsule separated from the rocket carrier (which remained in orbit) while the capsule, with Gagarin inside, started landing. It came down in Saratov, about 740 km (460 mi) east of Moscow.

④ **The cosmonaut** is separated from the chair at an altitude of 4,000 m (13,123 ft).

⑤ **The cosmonaut** lands in Saratov at 11:05.

**1961 HAM**
First chimpanzee sent into space on a suborbital trip.

**1961 VOSTOK 1**
In a 108-minute flight, the Russian Yuri Gagarin orbited Earth.

**1961 MERCURY**
15-minute suborbital flight by Alan Shepard, from NASA.

**1964 GEMINI 1**
Gemini I and II were launched unmanned in 1964 and 1965.

**1964 VOSKHOD 1**
It was the first time a three-person crew travelled into space.

**1965 VOSKHOD 2**
Aleksei Leonov took the first spacewalk.

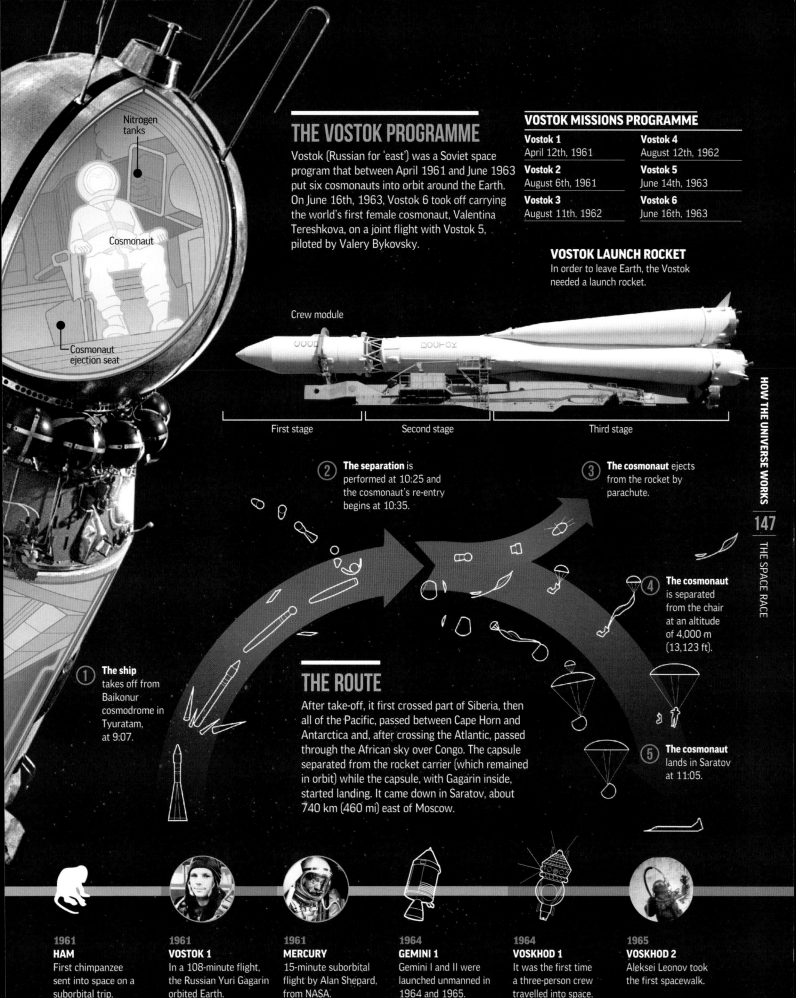

# UNITED STATES SPACECRAFT

**Between 1959 and 1963, the United States developed the Mercury program.** Before the first manned mission in May 1961, NASA sent three monkeys into space. The ships were launched into space with two rockets: Redstone for suborbital flights, and Atlas for orbital flights. The Little Joe was used to test the escape tower and the controls to abort the mission.

THRUSTERS

HEAT SHIELD

DOUBLE WALL

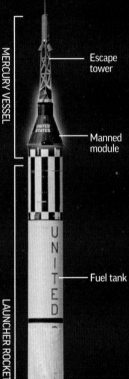

MERCURY VESSEL

Escape tower

Manned module

LAUNCHER ROCKET

Fuel tank

Oxidant tank

Engine

## THE MERCURY EXPERIENCE

After the launch of Sputnik I in 1957, and in the context of the Cold War, the United States was quick to start its own astronaut program. The development of the Mercury spacecraft was the impetus for starting the Apollo project, announced as "Lunar Flyby" in 1961 and then modified, by the wishes of President Kennedy, to reach the moon.

### FIRST TESTS
The first space flights were made by animals. Ham was the first monkey to fly in space. Equipped with sensors and remote controlled, Ham survived in space without problems.

### THE FIRST
On May 5th, 1961, Alan Shepard (1923-98) became the first American to fly aboard a Mercury spaceship. He then occupied important positions at NASA and in 1971 was part of the Apollo 14 mission.

### THE LAST
Gordon Cooper (1927-2004) was commander of the last Mercury mission, in May 1963, which lasted 22 orbits and closed the operational phase of the project. In 1965 he made a trip in the Gemini mission. He retired in 1970.

## 1,935 KG (4,266 LB)
**WEIGHT ON THE GROUND.**

### MERCURY

**First launch**
July 29th, 1960

**Maximum altitude**
282 km (175 mi)

**Diameter**
2 m (6 ft 6 in)

**Maximum duration**
22 orbits (34 hours)

**Organization**
NASA

**1965**
**MARINER 4**
Flew over Mars and took the first pictures of the planet.

**1965**
**GEMINI 3**
First manned flight of that programme.

**1965**
**DOCKING**
Gemini 6 and 7 managed to find each other and join up in space.

**1966**
**LUNA 9**
First landing on February 3rd. Photographs were sent to Earth.

**1966**
**SURVEYOR 1**
First American moon landing, on 2 June. More than 10,000 photos.

**1966**
**LUNA 10**
In April, the USSR sent another satellite, which sent radio signals.

① **Powered by the fuel,** the launcher takes off with the control module. The ship is fitted with three solid-fuel rockets.

Escape tower

Tower

Capsule

Booster engine

② **Once the escape tower** and booster engine have been jettisoned, the escape rockets ignite and the parachute system is armed.

Re-entry parachutes

③ **The capsule rotates 180 degrees.** Depending on the mission, between 1 and 22 orbits can be completed. Then it begins its descent.

Capsule

WINDOW FOR OBSERVATION

## FIRST JOURNEYS

The six successful Mercury missions were propelled by a solid-fuel rocket. The first flight, in May 1961, was a suborbital trip that lasted 15 minutes. Over the years the permitted time in space has increased due to better spacecraft.

Return

CONTROL PANEL

THRUSTERS

Main parachute

DRIVERS

PARACHUTE

④ **The descent begins** at an altitude of 6.4 km (4 mi). The capsule starts re-entry. The parachutes open.

AERODYNAMIC REGULATION

### MERCURY WITH ANIMALS

| | | |
|---|---|---|
| **Little Joe** | September 9th, 1959 | Sam |
| **Redstone** | January 31th, 1960 | Ham |
| **Atlas 5** | November 19th, 1959 | Enos |

### MERCURY WITH ASTRONAUTS

| | | |
|---|---|---|
| **Redstone 3** | May 5th, 1961 | Alan Shepard |
| **Redstone 4** | July 21th, 1961 | Gus Grissom |
| **Atlas 6** | February 20th, 1962 | John Glenn |
| **Atlas 7** | May 24th, 1962 | Scott Carpenter |
| **Atlas 8** | October 3rd, 1962 | Wally Schirra |
| **Atlas 9** | May 15th, 1963 | Gordon Cooper |

Landing in the ocean

RESCUE ROCKET

⑤ **Before rescue,** the pilot's parachute and the reserve parachute are released. They fall into the sea and are recovered.

# ONE GIANT LEAP FOR MANKIND

The space race culminated in Kennedy's words that pledged a landing before the end of the 1960s and the subsequent and successful arrival on the lunar surface. For the first time in history a man could walk on the Moon's surface, which including both the journey and the landing, lasted one week. It was the first journey that used two propulsion systems: one for take-off from Earth to the Moon and another to return from the Moon to Earth.

## TAKE-OFF

The module was powered by the Saturn V rocket, the heaviest ever built: almost 3,000 tons.

① **In two minutes and 42 seconds** the rocket reaches a speed of 9,800 km/h (6,090 mph) and enters Earth's orbit.

Launch platform

Stage 1

Gyro

② **The second stage** ignites and the ship reaches 23,000 km/h (14,290 mph).

Stage 3

The orbital and lunar modules stay together until the trajectory correction.

Linked modules

After reaching lunar orbit, the Eagle module separates and prepares its landing.

Correction

Module

## EAGLE LUNAR MODULE

It was divided into two parts, one for ascent and another for descent. It docked with the orbital module for the ascent and the descent.

## THE VOYAGE

The overall mission lasted about 200 hours. Two modules were used for the trip: one orbital (Columbia) and the other, the lunar module (Eagle). Both were attached to the Saturn V rocket until just after the third stage. After reaching lunar orbit, the Eagle module separated with two astronauts on board and prepared the landing. The return took place on July 24th. The stay on the moon lasted 21 hours and 38 minutes.

110 m (361 ft)

**Saturn V**
The rocket was as high as a 20-story building.

RADAR ANTENNA FOR COUPLING

CABIN

DRIVE CONTROL ASSEMBLY

EXIT PLATFORM

OXYGENATOR TANK

EQUIPMENT FOR EXPERIMENTS

### LM-5 EAGLE

| | |
|---|---|
| **Moon landing** | July 20th, 1969 |
| **Height** | 6.5 m (21 ft) |
| **Cabin volume** | 6.65 m³ (235 cu ft) |
| **Crew** | 2 |
| **Organization** | NASA |

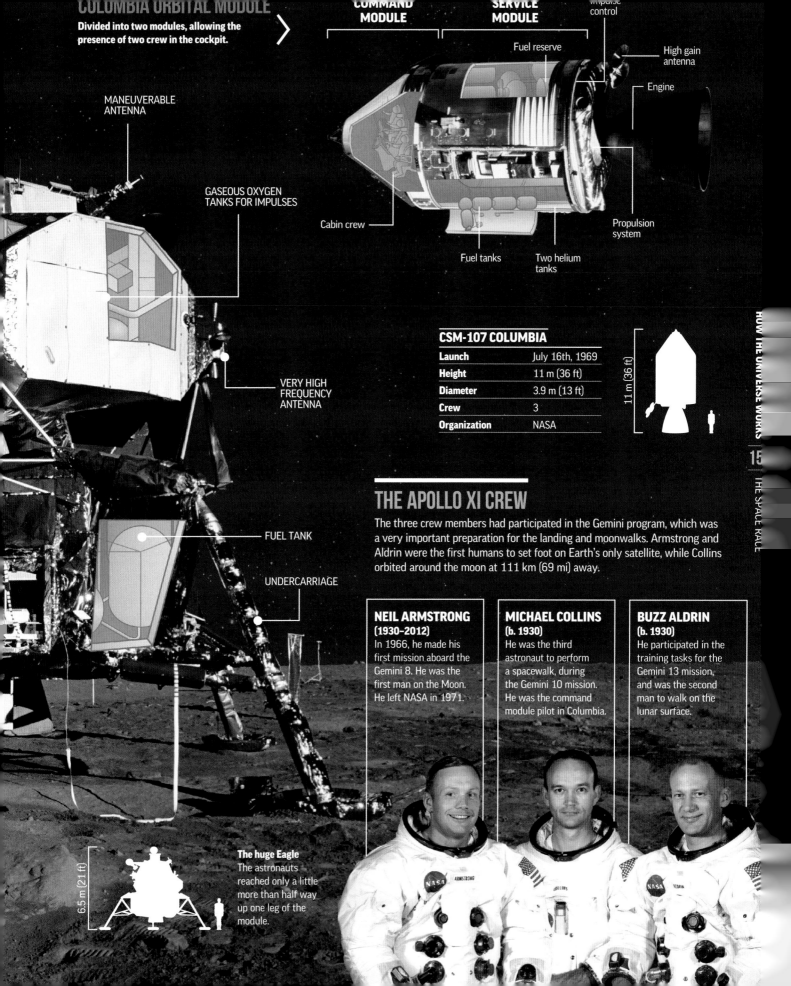

## COLUMBIA ORBITAL MODULE

Divided into two modules, allowing the presence of two crew in the cockpit.

**COMMAND MODULE**

**SERVICE MODULE**

Impulse control

Fuel reserve

High gain antenna

Engine

MANEUVERABLE ANTENNA

GASEOUS OXYGEN TANKS FOR IMPULSES

Cabin crew

Fuel tanks

Two helium tanks

Propulsion system

VERY HIGH FREQUENCY ANTENNA

FUEL TANK

UNDERCARRIAGE

### CSM-107 COLUMBIA

| Launch | July 16th, 1969 |
|---|---|
| Height | 11 m (36 ft) |
| Diameter | 3.9 m (13 ft) |
| Crew | 3 |
| Organization | NASA |

11 m (36 ft)

## THE APOLLO XI CREW

The three crew members had participated in the Gemini program, which was a very important preparation for the landing and moonwalks. Armstrong and Aldrin were the first humans to set foot on Earth's only satellite, while Collins orbited around the moon at 111 km (69 mi) away.

**NEIL ARMSTRONG (1930–2012)**
In 1966, he made his first mission aboard the Gemini 8. He was the first man on the Moon. He left NASA in 1971.

**MICHAEL COLLINS (b. 1930)**
He was the third astronaut to perform a spacewalk, during the Gemini 10 mission. He was the command module pilot in Columbia.

**BUZZ ALDRIN (b. 1930)**
He participated in the training tasks for the Gemini 13 mission, and was the second man to walk on the lunar surface.

**The huge Eagle**
The astronauts reached only a little more than half way up one leg of the module.

6.5 m (21 ft)

# THE APOLLO PROGRAM

**Six Apollo missions managed to land on the surface of the Moon, the exception was Apollo 13, which after one of its oxygen tanks exploded, managed to return to Earth.**
They all meant that the Moon ceased to be something unattainable. Each of these trips, in addition to providing data, drove the development of space science and increased the desire to carry out other expeditions to different parts of the Solar System.

## THE APOLLO MISSIONS

The Apollo program began in May 1961. It was one of the greatest triumphs of modern technology. Six expeditions were able to land on the surface (Apollo 11, 12, 14, 15, 16 and 17). From the 24 astronauts who travelled there, 12 walked on the Moon. The Apollo Lunar Module was the first spacecraft designed to fly in a vacuum without an aerodynamic design.

### TWENTY-ONE CHOSEN ONES
Apollo took six expeditions to the moon with a total of 24 astronauts.

### LUNAR MATERIAL
Moon rock samples turned out to be similar to the Earth's mantle.

### ROUTE
This is the total distance travelled by the Lunar Rover on Apollo XV, XVI and XVII.

### DURATION
The Apollo XVII mission, the longest of all, lasted almost 302 hours.

### END OF MISSION
The Apollo-Soyuz mission ended the lunar space race.

## 336 KG
### (741 LB)

## 25 KM
### (16 MILES)

## 301:51':50"

## THE LUNAR ROVER

**Electric vehicle used by the astronauts to explore the lunar surface.**

HIGH GAIN ANTENNA

LOW GAIN ANTENNA

TELEVISION CAMERA

TELEVISION CAMERA

MANUAL CONTROLLER

COMMUNICATION TRANSMISSION UNIT

DATA CONSOLE

### LUNAR ROVER

| Launch | July 1971 |
|---|---|
| Length | 3.1 m (10 ft) |
| Width | 1.14 m (3.74 ft) |
| Velocity | 16 km/h (10 mph) |
| Organization | NASA |

## APOLLO MISSIONS

### 1970
### APOLLO 13
The explosion in the liquid oxygen tank in the service module caused the early return of the crew, made up of James Lovell, Fred Haise and Jack Swigert.

### 1972
### SAMPLES
In the last mission, the Apollo 17 astronauts Eugene Cernan and Harrison Schmitt travelled around the Moon in the Lunar Rover and took rock samples from the surface.

### 1975
### APOLLO-SOYUZ
The Apollo and Soviet Soyuz spacecraft performed a docking in space, in the first joint mission between NASA and the Soviet Space Agency. It was the last Apollo mission.

# THE LUNAR ORBITER

The Lunar Prospector was launched in 1998 and was in space for 19 months. It orbited the Moon at an altitude of 100 km (62 mi) travelling at a speed of 5,500 km/h (3,420 mph), and completed the orbit every two hours. Its purpose was to map the surface composition and possibly to recognize water deposits in the form of ice and to measure the Moon's magnetic and gravitational fields.

**LUNAR PROSPECTOR**
It consists of a cylinder covered with thousands of photovoltaic panels.

**Antenna**
For communication with Earth.

**Gamma -ray spectrometer**
Searches for the existence of potassium, oxygen, uranium, aluminium, silicon, calcium, magnesium and titanium.

**Thrusters**

Solar panel

**Magnetometer**
Finds the magnetic fields in the vicinity of the ship.

**Alpina particle spectrometer**
Detects particles emitted by radioactive gases.

**Neutron spectrometer**
Detects neutrons from the lunar surface.

| LUNAR PROSPECTOR | |
| --- | --- |
| Launch | January 1998 |
| Flight to Moon | 105 hours |
| Weight | 295 kg (650 lb) |
| Cost | US $63 million |
| Organization | NASA |

COLLECTION BAG FOR SAMPLES

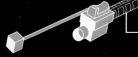

**209 KG**
**(460 LB)**
WEIGHT ON EARTH

**35 KG**
**(77LB)**
WEIGHT ON THE MOON

## END OF THE APOLLO PROGRAM

After six lunar landings, the Apollo program was terminated. While Apollo 18, 19 and 20 were cancelled due to budget issues, the program put the United States at the head of the space race.

### JAMES A. LOVELL, JR
(b. 1928)
Pilot of Apollo 13. The mission was aborted because of the explosion of a service module on-board. Lowell was also an emergency pilot on Gemini 4 and pilot on Gemini 7 and 12.

### HARRISON SCHMITT
(b. 1935)
American geologist. He travelled aboard the spacecraft Apollo 17, the last Apollo mission, and was the first geologist and the only civilian to work on the Moon.

### ALEKSEI LEONOV
(b. 1934)
Russian cosmonaut. He was part of the test project for Apollo and Soyuz docking which lasted seven days. Aboard the Voskhod II, he was the first person to walk in space.

## LATER MISSIONS

**1994**
### CLEMENTINE
The Clementine spacecraft orbited the Moon and mapped its surface. It also transmitted radio signals into the shadowed craters near its South Pole.

**2003**
### SMART
ESA launched the Smart I, its first unmanned spacecraft bound for the Moon. Its aim was to analyze unexplored regions and test new technologies.

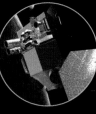

**2009**
### LRO
NASA launched a rocket carrying the Lunar Reconnaissance Orbiter spacecraft in order to search for ice in the polar regions of the Moon.

# ECHOES
# OF THE PAST

**Thanks to the data sent in 2001 from the NASA observatory WMAP (Wilkinson Microwave Anisotropy Probe),** scientists have managed to make the first detailed maps of cosmic background radiation. Cosmic background radiation is thought to be the echoes of the Big Bang. Experts believe that this map reveals clues about when the first generation of stars was formed.

## THE WMAP MISSION

The observatory observes the entire sky every six months for two years to ensure that data is accurate. It then compares the obtained maps to check for consistency.

## 1,851 LB
### (840 KG)

**WEIGHT OF THE OBSERVATORY WHEN ON THE GROUND.**

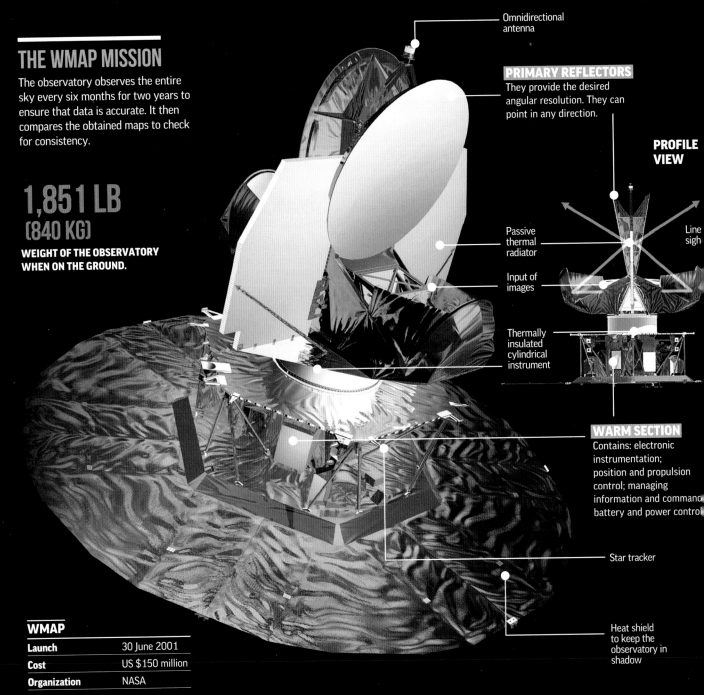

Omnidirectional antenna

**PRIMARY REFLECTORS**
They provide the desired angular resolution. They can point in any direction.

**PROFILE VIEW**

Line sigh

Passive thermal radiator

Input of images

Thermally insulated cylindrical instrument

**WARM SECTION**
Contains: electronic instrumentation; position and propulsion control; managing information and command battery and power control

Star tracker

Heat shield to keep the observatory in shadow

| WMAP | |
|------|------|
| **Launch** | 30 June 2001 |
| **Cost** | US $150 million |
| **Organization** | NASA |

# OBSERVATION

In order to observe the whole sky, the probe is located at the so-called L2 Lagrange Point, 1.5 million km (0.9 million mi) from Earth. This point provides a stable environment, away from the influence of the sun. WMAP observes the sky at different stages and measures temperature differences between different cosmic regions. Every six months it completes a full sky coverage.

② 
## DAY 90

**(3 MONTHS)**
**The probe has completed coverage of half the sky. Each hour it covers a sector of 22.5°.**

**WMAP TRAJECTORY**
Before heading to the L2 Lagrange Point, the probe performed a flyby of the moon, using lunar gravity to propel it towards L2.

③ 
## DAY 180

**(6 MONTHS)**
**It has completed one full sky view. The process is repeated four more times.**

① 
## DAY 1

**Thanks to its ability to focus in two directions simultaneously, WMAP is able to observe a large area daily.**

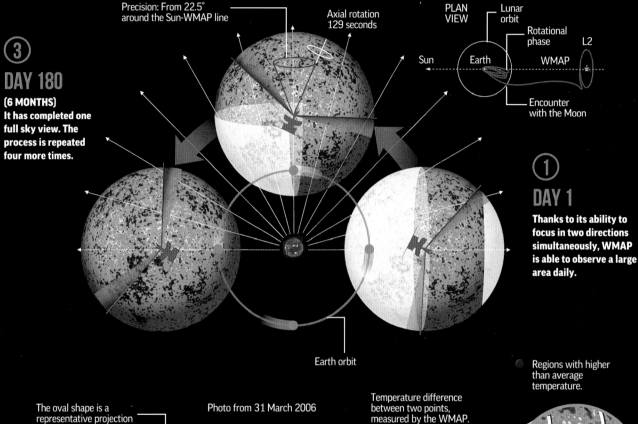

Precision: From 22.5° around the Sun-WMAP line

Axial rotation 129 seconds

PLAN VIEW

Lunar orbit

Rotational phase

L2

Sun

Earth

WMAP

Encounter with the Moon

Earth orbit

Regions with higher than average temperature.

Regions with lower than average temperature.

The oval shape is a representative projection to display the whole sky.

Photo from 31 March 2006

Temperature difference between two points, measured by the WMAP.

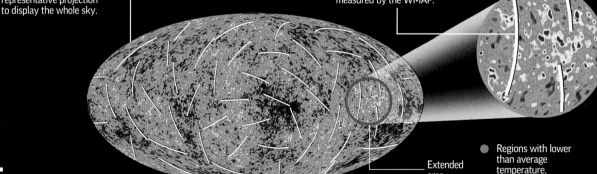

Extended area

# MAP

The different colors of the regions detailed in the WMAP sky map are very slight temperature differences in the cosmic microwave background. This radiation, remnants of the Big Bang, was discovered 40 years ago, but can only now be described in detail.

**COBE, THE PREDECESSOR**
COBE's results from 1989 provided the kick-start for the WMAP project. The resolution was much lower, so the spots are larger.

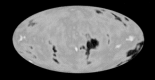

**WORKING IN SPACE**
NASA astronaut Donald R. Pettit,
hanging from the International
Space Station during a session of
extravehicular activity.

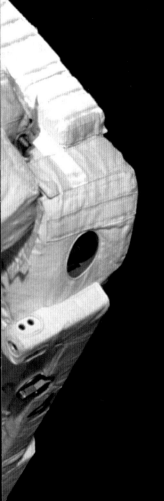

## CHAPTER 6

# EXPLORATION
# MISSIONS

The search for answers about the origin and structure of the universe has led man to send increasingly sophisticated vehicles into space, as well as other unmanned reconnaissance artifacts such as space probes and telescopes. In addition, there are space stations permanently inhabited by astronauts.

In the last fifty years, all the planets of the Solar System have been visited by space probes, including Uranus and Neptune, the most distant planets. In some cases, the visit was only a flyby mission, which nevertheless provided data impossible to obtain from the Earth. Other missions have involved placing space probes in orbit around a planet. Yet other missions have landed probes on Venus, Mars, and Titan (one of Saturn's moons). In 1969, humans succeeded in walking on the Moon, and there are now plans to send humans to the planet Mars.

## UNMANNED SPACECRAFT

All planetary missions have been accomplished with unmanned spacecraft. When possible their voyages have taken advantage of the gravitational field of one or more planets in order to minimize fuel requirements.

International Space Station

### EARTH

Many artificial satellites and manned missions have orbited and continue to orbit the Earth. The orbiting International Space Station always has a crew onboard.

Space Shuttle

### MOON

The Apollo missions (1969-72) took a total of 12 astronauts to the surface of the Moon. They are the only missions that have taken humans beyond the Earth's orbit. The United States, Europe, Japan, Russia and China are preparing new manned missions to the Moon.

### MERCURY

Visited in 1974-75 by Mariner X on three flybys, with a closest approach of 327 km (203 mi). The probe mapped 45 percent of the planet and made various types of measurements. In 2011, the probe Messenger went into orbit around Mercury after making flybys in 2008 and 2009.

### VENUS

The most visited celestial body after the Moon, Venus has been studied by orbiting spacecraft and by landers, many in the 1970s and 1980s. During the Vega and Venera missions and the Mariner and Magellan missions, the surface of the planet was mapped and even excavated, and the atmosphere was analyzed. Between 2005 and 2014, the spacecraft Venus Express studied the planet from orbit.

**DISTANCE FROM THE SUN**

| Mercury | Venus | Earth | Mars |
|---------|-------|-------|------|
| 57,900,000 km (36,000,000 mi) | 108,000,000 km (67,000,000 mi) | 150,000,000 km (93,000,000 mi) | 227,900,000 km (141,600,000 mi) |

## JUPITER

The giant of the Solar System was visited for the first time by Pioneer 10 in 1973. Another seven spacecraft (Pioneer 11, Voyagers 1 and 2, Ulysses, Cassini, Galileo, and New Horizons) have made flybys of the planet since then. Galileo studied Jupiter and its moons for eight years from 1995 to 2003, and it transmitted images and data of incalculable scientific value.

## NEPTUNE

The distant blue giant has been visited only once, in 1989, by Voyager 2.

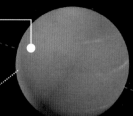

## URANUS

In 1986, Uranus was visited by Voyager II, which took photographs and readings of the planet. It is the only mission that has reached Uranus.

# 7 YEARS

**THE TIME IT TOOK FOR THE CASSINI PROBE TO TRAVEL FROM THE EARTH AS FAR AS JUPITER. GALILEO REACHED JUPITER IN SIX YEARS.**

## SATURN

Only four missions have visited Saturn. The first three—Pioneer 11 (1979), Voyager 1 (1980), and Voyager 2 (1981)—flew by at distances of 34,000 to 350,000 km (21,000 to 220,000 mi) from the planet. Cassini, in contrast, was placed in orbit around Saturn in 2004, and it has obtained amazing images of the planet and its rings. Part of the Cassini mission was to launch the Huygens probe, which successfully landed on the surface of Saturn's mysterious moon Titan.

# BEYOND THE SOLAR SYSTEM

Having left behind the orbit of Neptune, the space probes Pioneer 10 and 11 and Voyager 1 and 2 are bound for the edge of the Solar System.

# EROS

**IN 2000, THE PROBE ENTERED ORBIT AROUND THE ASTEROID 433 EROS.**

## PIONEER 10 AND 11

They were launched in 1972 and 1973 and visited Jupiter and Saturn. Contact with the probes was lost in 1997 and 1995, respectively. They carry a plaque with information about the Earth and human beings in anticipation that they may eventually be found by an extraterrestrial civilization.

## VOYAGER 1 AND 2

Launched in 1977, they carry a gold-plated disk with music, greetings in various languages, sounds and photographs from the Earth, and scientific explanations. The probes passed Jupiter, Saturn, Uranus, and Neptune and remain in contact with the Earth.

## MARS

In 1965, Mariner 4 took the first 22 close-up images of Mars. Since then the planet has been visited by many orbiters and by probes that have landed on its surface. Among the most noteworthy are the missions of Viking (1976), Mars Pathfinder (1997), Mars Exploration Rovers (2004), and the Mars Science Laboratory (2011).

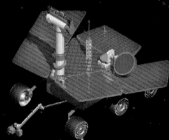

Mars Exploration Rover (2004)

| **Jupiter** | **Saturn** | **Uranus** | **Neptune** |
|---|---|---|---|
| 778,000,000 km (483,000,000 mi) | 1,427,000,000 km (887,000,000 mi) | 2,870,000,000 km (1,780,000,000 mi) | 4,500,000,000 km (2,800,000,000 mi) |

# POINT OF DEPARTURE

The launch bases for space rockets are often located in regions near the equator to facilitate the launching of spacecraft. It is also better if it is a coastal site, to facilitate the transport of materials, and it should have low population density to minimize damage in case of accidents during launch. One of these bases is at Cape Canaveral, Florida, in the United States.

## TERRESTRIAL PLATFORM

This steel giant is the point from which spacecraft take off. It is made up of fixed and rotating structures. The orbiter is transported from the assembly building to the terrestrial platform on a caterpillar-carrier platform.

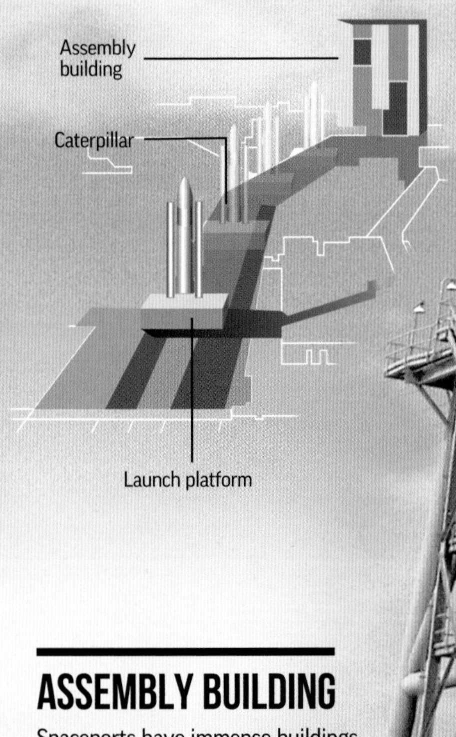

Assembly building

Caterpillar

Launch platform

## ASSEMBLY BUILDING

Spaceports have immense buildings where you prepare and assemble the rocket boosters and external tank with the shuttle. The dimensions of these working hangars are amazing: they are 160 m (525 ft) high, 218 m (715 ft) long and 118 m (387 ft) wide.

**ROTATING SERVICE STRUCTURE**
It has a height of 57.6 m (189 ft) and moves in a semicircular path around the shuttle.

**LIFT**
Astronauts begin their isolation in the lift. From here they go to the white room, and then into the shuttle.

## LIGHTNING ROD

Protects people, the shuttle and other elements of the platform against electrical shock. It is 106 m (348 ft) high.

## FIXED SERVICE STRUCTURE

At 75 m (246 ft) high, it is spread over 12 floors. It has three arms that connect to the shuttle.

BOOSTER ROCKETS

## WHITE ROOM

Exclusively for astronauts. From here they go on to the shuttle.

ORBITER ACCESS ARM

USA

NASA
Endeavour

# FLOATING PLATFORM

Some countries have developed projects for floating launch platforms, from which it is easier and safer to locate to the terrestrial equator. This is the place where the Earth's rotation speed is highest, which favors launching space missions into orbit.

Rocket

Platform

1. **Assembly**
A rocket is assembled in a mounting ship, 200 m (656 ft) long.

2. **Transfer**
The rocket is transferred to the launch pad.

3. **Storage**
The rocket is stored until the launch. The mounting boat floats away.

## OTHER LAUNCH BASES

Spaceports are located preferably in close proximity to the equator because vehicles launched eastwards from anywhere on this line travel with increased efficiency in terms of speed, cost and payload capacity.

### FIRST LAUNCHES FROM MAJOR BASES

Plesetsk
(1966)

Kennedy
(1967)

EQUATOR

Kourou
(1970)

San Marco
(1967)

## REAR SERVICE MAST

These structures connect the platform with the ship. They provide oxygen and hydrogen to the external tank.

40 m (131 ft)

## SHUTTLE CRAWLER

The orbiter moves to the platform on twin caterpillar tracks. A laser system guides it accurately at a speed of 3.2 km/h (2 mph).

# ROCKETS

Developed in the first half of the 20th century, rockets are necessary for sending any kind of object into space. They produce enough force to leave the ground with cargo and, in a short period of time, they acquire the speed necessary to escape gravity completely and orbit around the Earth. Approximately, more than one rocket per week is launched into space from somewhere in the world.

## ARIANE 5

**First successful launch**
October 30th, 1997

**Diameter**
5 m (16 ft)

**Total height**
51 m (167 ft)

**Weight of boosters**
277 tons each (full)

**Cost of the project**
7 billion Euros

**Maximum load**
6,200 kg (13,670 lb)

**Organization**
ESA

## SPACEFLIGHTS

Access to space—whether for placing satellites into orbit, sending probes to other planets, or launching astronauts into space—has become almost routine and is big business for countries that have launch capabilities.

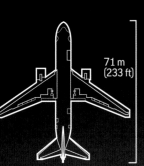

71 m (233 ft)

51 m (167 ft)

**ARIANE 5**

17 m (56 ft)

**BOEING AIRCRAFT**

**SPACE SHUTTLE**

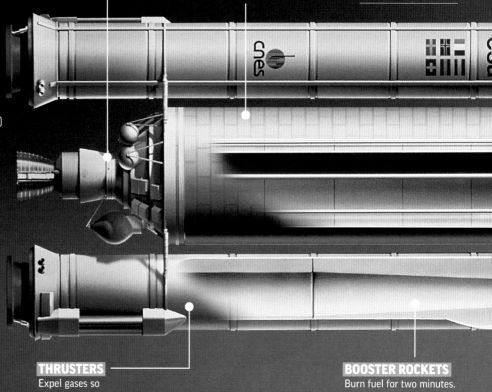

**MAIN ENGINE**
Burns for 10 minutes.

**THERMAL INSULATION**
To protect the combustion chamber from the high temperatures of the burning fuel, the walls are sprayed with rocket fuel. This process manages to cool the engine off.

**THRUSTERS**
Expel gases so that the rocket can begin its ascent.

**BOOSTER ROCKETS**
Burn fuel for two minutes.

# 40,000 KM/H
## (24,855 MPH)
**LIFT-OFF VELOCITY.**

# 746,000 KG
## (1,645,000 LB)
**WEIGHT ON THE GROUND.**

## OPERATION OF THE ENGINE

Before take-off, fuel ignition is started. The main engine turns on and only if the ignition is successful the thrusters are turned on. The rocket takes off and two minutes later the thrusters are turned off when they are out of fuel. The engine stays on for a few minutes and then switches off. A small engine puts the satellite into orbit.

**MOTOR**

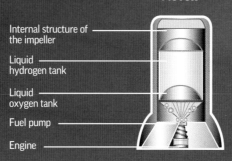

Internal structure of the impeller

Liquid hydrogen tank

Liquid oxygen tank

Fuel pump

Engine

**COMPONENT PARTS**

Loading system

Guidance system

Propulsion system

## CHEMICAL ROCKETS ACCORDING TO FUEL

In liquid rockets, the hydrogen and oxygen are in separate containers. In solid rockets, they are mixed and placed in a single cylinder.

Gases removed

**LIQUID**  **SOLID**  **HYBRID**

## THERMAL INSULATION

To protect the combustion chamber from the high temperatures of the fuel burned, the walls are covered by the used rocket propellant. This cools the engine.

Cover

Propellant

Insulation

## LIQUID HYDROGEN TANK
Contains 225 tons.

## LIQUID OXYGEN TANK
Contains 130,000 kg (286,000 pounds) for combustion.

## UPPER PAYLOAD
Carries up to two satellites.

## UPPER ENGINES
Release the satellites at a precise angle and speed.

## LOWER PAYLOAD
Carries up to two satellites.

## CONICAL NOSE CONE
Protects the cargo.

## HOW IT WORKS

To do its job, the rocket must overcome gravity. As it rises, the mass of the rocket is reduced through the burning of its fuel. Moreover, as the distance from the Earth increases, the effect of gravity decreases.

## ACTION AND REACTION
The thrust of the rocket is the reaction resulting from the action of the hot exhaust escaping from the rocket.

Rocket's thrust

Earth's gravity

## TYPE OF ROCKET ACCORDING TO PROPULSION

The chemical propulsion rocket is the most widely used. It is driven by combustion. The nuclear type is driven by fission or fusion. The ion motor provides the possibility of electrically charging atoms by stripping electrons.

Thrust

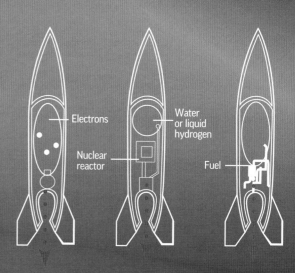

Electrons

Water or liquid hydrogen

Nuclear reactor

Fuel

**ION**  **NUCLEAR**  **CHEMICAL**

# TAKEOFF CHRONICLE

Although just 50 years have passed since the first space flights to today, access to space has become almost routine. The amazing thing about space rockets, those colossal machines that involve monstrous energy levels, is perhaps that the technology, despite advances in computers, engines and guidance systems in the second half of the twentieth century, has hardly changed from its foundations.

## LAUNCH SEQUENCE

The countdown for the Ariane V typically lasts six hours. At the end of the countdown, the launch begins with the ignition of the main stage's liquid-fuel engine. Seven seconds later the two solid-fuel boosters are ignited. If there are problems before the boosters are ignited, the launch can be aborted by shutting down the main stage.

**06:00:00**
Start of the sequence.

**04:30:00**
The tank starts to be filled.

**01:00:00**
Mechanical reinforcements.

**00:06:30**
Start of the synchronized sequence.

**00:00:00**
The liquid fuel engines on the main stage light.

### HOW IT IS ORIENTATED

The rocket's guidance computer uses data from laser gyroscopes to control the inclination of the nozzles, directing the rocket along its proper flight path.

Laser gyroscope

Electrical signals

Computer

Cardan gimbal joints

Nozzle inclinations

### ① FIRST STAGE

**The solid-fuel boosters are ignited. The rocket begins to lift off 0.3 second later.**

### ② DETACHMENT

**At 60,000 m (200,000 feet) the solid-fuel boosters separate and fall to the ocean in a secure area.**

### FAIRING

They separate as the atmosphere becomes sparse, and therefore poses no risk to the payload.

### SOLID-FUEL BOOSTERS

Provide 90 percent of the initial thrust needed to launch the Ariane V. The boosters are 31 m (102 feet) high and contain 238,000 kg (525,000 pounds) of fuel.

### EXPLOSIVE BOLTS

Separate the boosters from the main stage and the main stage from the second stage.

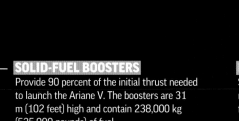

# 110.6 M
## (363 FT)

**THE HEIGHT OF SATURN V, THE
LARGEST ROCKET EVER LAUNCHED,
THAT TOOK MAN TO THE MOON.**

## ④ FINAL PHASE

**The upper-stage rocket is the only
rocket not used on the launching pad.
Instead it is used to insert the payload
into its proper orbit. The rocket can
be reignited after it is shut down and
can burn for a total of 19 minutes.**

### SECOND PHASE

Separates at an altitude of
about 120 km (75 mi) and
falls to Earth.

## ③ THE MAIN STAGE

**The main stage, ignited at
the end of the countdown,
separates and falls back to
Earth. Its supply of liquid
hydrogen and oxygen has
been used up.**

### STAGES OF THE ROCKET

It has two solid propellant motors. To get started
it needs fuel storage and fuel elevators.

**Final stage**: contains
the payload and ignites
once the rocket is
already in space.

**Second stage**: this is
the main stage.

**First stage**: these are
the solid fuel rockets.

## STAGES OF DETACHMENT

The Ariane V rocket has three stages.
On the launch pad the first two stages
are lit. In the ascent and as each section
is consumed, they separate from the
spacecraft through a series of explosive
charges placed in the first and second
stage. In the third stage the control
elements and the cargo hold are housed.

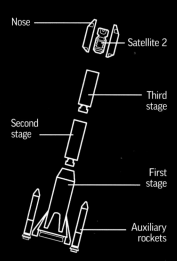

Nose

Satellite 2

Second
stage

Third
stage

First
stage

Auxiliary
rockets

## LAUNCH WINDOW

Rockets must be launched at predetermined times, which depend on the
objective of the launch. If the objective is to place a satellite into orbit, the
latitude of the launched rocket needs to coincide with the trajectory of the
desired orbit. When the mission involves docking with another object in
space, the launch window might fall within only a few minutes.

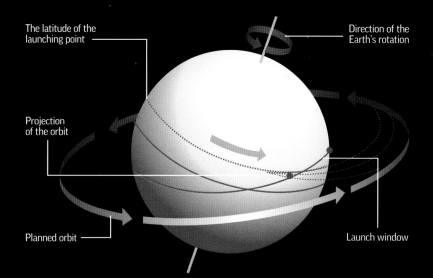

The latitude of the
launching point

Direction of the
Earth's rotation

Projection
of the orbit

Planned orbit

Launch window

# SPACE SHUTTLE

**Unlike conventional rockets, the space shuttles were used over and over again to put satellites into orbit.** Until 2011, these vehicles were used to launch and repair satellites and as astronomical laboratories. The American fleet had five space shuttles over its history: Challenger and Columbia (exploded in 1986 and 2003, respectively), Discovery, Atlantis, and Endeavour. The space shuttle program ended in 2011 with the retirement of the three remaining shuttles.

## REUSABLE

The space shuttle was the first spacecraft capable of returning to Earth on its own being used in multiple missions. The orbiter played a key role in building the International Space Station.

## CABIN

Divided into two levels: an upper one for the pilot and co-pilot (and up to two astronauts), and a lower one where everyday work is done. The habitable volume of the cabin is 2,472 ft³ (70 m³).

**Controls**
In the cockpit there are more than 2,000 separate controls.

Control keypad

Co-pilot's seat

Pilot's seat

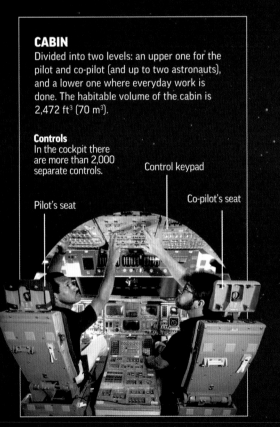

## SATELLITE
Stays in the cargo hold and is moved by the arm.

## MECHANICAL ARM
Moves satellites in and out of the cargo module.

## ① ORBITER

**The orbiter carried the crew and the load (usually satellites).**

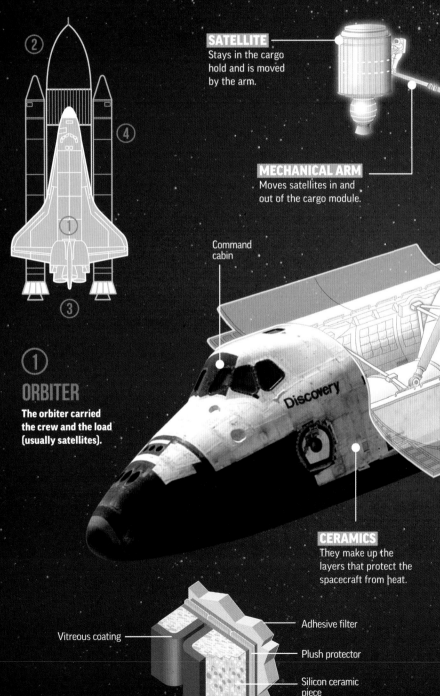

Command cabin

Discovery

## CERAMICS
They make up the layers that protect the spacecraft from heat.

Vitreous coating

Adhesive filter

Plush protector

Silicon ceramic piece

## EXTERNAL FUEL TANK

**②**

Connects the shuttle to the launcher rockets. Carries loads of liquid oxygen and hydrogen, which are combusted through a tube connecting each container with the next. The tank is lost on each trip.

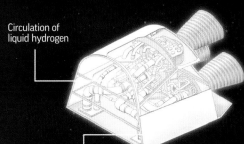

Liquid oxygen

Liquid hydrogen

## THE MAIN ENGINES

**③**

There are three of these, which feed liquid oxygen and hydrogen from the external tank. Each engine has a controller based on a digital computer, which makes adjustments to the thrust and corrects the fuel mixture.

Circulation of liquid hydrogen

Heat shield

## VERTICAL WING

The vertical wing was used during the descent to stabilize and to control the direction.

## ORBITAL MOTORS

Provide the thrust for entering orbit and orbital adjustments that may be needed. They are located on the outside of the fuselage.

United States

## DELTA WINGS

The shuttle had to descend gliding without fuel. So its wings resembled those of a paper airplane.

Ignition section

## SOLID ROCKETS

**④**

They are designed to last about 20 flights. After each trip they are retrieved from the ocean and refurbished. They take the shuttle to an altitude of 44 km (27 mi) and can support the full weight of the shuttle.

Solid fuel

Thruster mouth

## GATES

They open when the device reaches low Earth orbit. They are thermal panels that protect the spacecraft from overheating.

## THERMAL PROTECTION

When a shuttle re-enters Earth's atmosphere, the friction heats the surface to temperatures of between 300° and 1,500° C (572 to 2,732° F). To avoid melting, the spaceship must have protective layers.

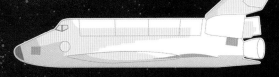

**Felt**. The heat is less than 370° C (698° F).

**Silicon ceramics**. 370–648° C (698 to 1,198° F).

648–1260° C (1,198 to 2,300° F). Also silicone.

**Metal or glass**, no thermal protection.

**Carbon** in areas above 1,260° C (2,300° F).

# PROFESSION: ASTRONAUT

Before embarking on a mission in space, candidates must undergo rigorous tests as the tasks they will have to perform are delicate and risky.

They should study mathematics, meteorology, astronomy and physics intensively, and acquire familiarity with computers and space navigation. They should also do physical exercise to help adjust to the low gravity in orbit and still be able to carry out repair work.

## UNIT FOR MANNED MANEUVERING

The training program is difficult and exhausting. Every day there are activities in flight simulators and simulations with specialized computers.

**CAMERA**
Color television equipment.

**COMPUTER**
Pocket communications equipment.

Visor

Digital camera

Image controller

**OXYGEN**
Enters through this part of the suit.

**COOLANT**
Provides a thermal layer and protection from meteorites.

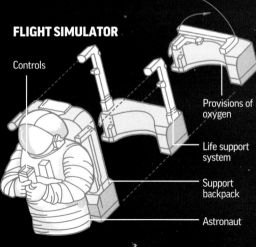

### FLIGHT SIMULATOR

Controls

Provisions of oxygen

Life support system

Support backpack

Astronaut

**1965**
**Astronaut Edward White** used this spacesuit to make a spacewalk in the area near the Gemini rocket.

**1969**
**Neil Armstrong** wore this suit when he performed the historic first spacewalk on the surface of the Moon.

**1984**
**Bruce McCandless** conducted the first spacewalk without being attached to the shuttle in this suit.

**1994**
**Space shuttle** astronauts had suits that are much more modern as well as reusable.

### THE HELMET
Contains a microphone for the communications equipment.

Treated plastic helmet to prevent fogging.

Snoopy helmet.

Microphone

### VISOR
For protection from the Sun.

### ORIFICE
For entry and exit of water.

### BELT
Holds the astronaut down to manage zero gravity.

### GLOVES
To protect the astronaut's hands.

### PARTS OF THE SUITS
The fabrics that compose the suit are specially designed to protect the astronaut's body.

FABRIC WITH WATER -CARRYING TUBES

NYLON

NEOPRENE

THERMAL COVER AGAINST MICROMETEORITES

### RESCUE SPHERE
They serve to help crew with no suits. They are made of spacesuit materials and have a reserve of oxygen.

OXYGEN RESERVE

EXTERIOR

CARRYING HANDLE

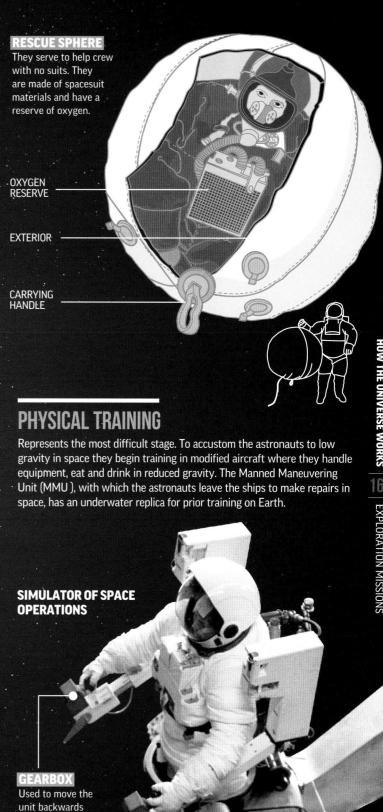

## PHYSICAL TRAINING
Represents the most difficult stage. To accustom the astronauts to low gravity in space they begin training in modified aircraft where they handle equipment, eat and drink in reduced gravity. The Manned Maneuvering Unit (MMU ), with which the astronauts leave the ships to make repairs in space, has an underwater replica for prior training on Earth.

### SIMULATOR OF SPACE OPERATIONS

### GEARBOX
Used to move the unit backwards and forwards.

### CONTROL PEDAL
Support for the astronaut.

# FAR FROM HOME

Leaving the Earth to live in a space station or make a shuttle journey implies adjusting to environments that man is not used to: no water, no air pressure and an absence of oxygen. Everything must be provided on board: water is made electrically from oxygen and hydrogen, salt for food is liquid and waste is powdered.

## LIVING AREA

The Space Shuttle living module was located at the tip. On the upper level you found the cockpit, and on lower levels, compartments for sleeping and living, and the hatch.

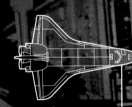

Cabin

Dormitory

Bathroom

Wardrobe

Hatch

## HOME FROM HOME

Space stations serve as the astronauts' home for weeks or even months. This is a prototype of the space station's module.

## PHYSICAL EFFECTS

Life in space can cause harmful effects on different body systems on returning to Earth. In many situations, living in confined spaces can cause psychological damage. In addition, the radiation emitted by the Sun can cause severe damage.

Diseased bone

Healthy bone

Hallucinations and dizziness

Respiratory system

Circulatory system

Muscular system

## CALCIUM LOSS IN THE BONES

In microgravity, bone tissue is not regenerated, but absorbed. The missing mass may appear as excess calcium in other body parts (e.g., kidney stones).

# 90 MINUTES
**DURATION OF THE DAY IN ORBIT**

Sleeping bag

## ① SLEEP

**ONCE A DAY**
In one space day the Sun rises and sets every hour and a half. Astronauts try to sleep eight hours a day, once at the end of each 'Earth day'. They have to sleep tied up so they do not float away.

## ② CLEANLINESS

They all wear the same clothes. After bathing, they change, because there is no way to do laundry in space. To go to the bathroom they use an air suction system, because it is impossible to use water.

## ③ FOOD

**THREE TIMES A DAY**
During the day, the astronauts have breakfast, lunch and dinner. They have to be very careful putting the food in their mouths, and they have to drink plenty of water because they can suffer dehydration.

## ④ WORK

**EIGHT HOURS A DAY**
They work four hours on Saturdays, and Sundays are free days. During the week there is a normal workday. The most commonly performed tasks are maintenance and scientific experiments.

## ⑤ EXERCISE

**TWO HOURS A DAY**
To maintain their health, astronauts must do physical exercise every day. As muscle is lost due to weightlessness, exercise helps maintain muscle tone.

Device for toning muscles

Spacesuit for work in space

# 72
**VARIETY OF MEALS**

# 20
**VARIETY OF DRINKS**

# SPACE STATIONS

**Living on space stations makes it possible to study the effects of remaining in outer space for extended periods of time,** while providing an environment for scientists to conduct experiments in laboratories. These stations are equipped with systems that provide the crew with oxygen and that filter exhaled carbon dioxide.

## 450 TONS
**APPROXIMATE WEIGHT ON EARTH.**

Photovoltaic panels

CAM (Centrifuge Accommodation Module)

Columbus laboratory

Harmony module

Space shuttle

## THE ISS

The International Space Station (ISS) is the result of the merger of NASA's Freedom project with Mir-2, run by the Russian Federal Space Agency (RKA). Construction began in 1998 and it continues to expand, using modules provided by countries across the globe. Its inhabitable surface area is equal to that of a Boeing 747.

### SIZE

Weight:
415 tons

20 m

51 m

108 m

## ORBIT

**THE ISS PERFORMS AROUND 16 COMPLETE ORBITS OF THE EARTH EACH DAY, AT A HEIGHT OF BETWEEN 335 TO 460 KM (208 TO 286 MI).**

### ORBIT

**Height:**
380 km

**Orbital period:**
One turn around the world every 91.34 minutes

**Inclination:**
15.76 degrees

**Average speed:**
27,743 km/h

## STAGES OF CONSTRUCTION

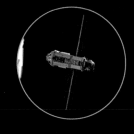

### NOVEMBER 1998
**Zarya Module**
First section put into orbit. Provided energy in the first stages of ISS assembly. In December, the Unity Module connected passage between the living and work area modules. Contributed by the European Union.

### JULY 2000
**Zvezda Module**
The structural and functional center of the ISS. Entirely built and placed into orbit by the Russians. In November, the structural module P6 Truss incorporates radiators for dissipating the heat generated in the station.

### FEBRUARY 2001
**Laboratory Destiny**
Holds 24 equipment racks. This is where scientific experiments in microgravity environments are performed. In November 2002, the P1 Truss was added opposite the S1 Truss as part of the integrated truss assembly. The truss radiator panels protect the ISS from the extreme temperatures.

Wardrobe

Beds

Control and
communications area

Destiny
laboratory

SPP (Science
Power Platform)

Unfolding
solar panels

## ZVEZDA MODULE

Built by Russia and assembled
in 2000. An example of a
module that provides vital
support and accommodation
for the crew.

ATV (Automated
Transfer Vehicle)

Zarya
module

Thermal
control
panels

## COMPOSITION

② **Module**
The arms gets closer
to the module,
in preparation to
couple to Zarya.

③ **Union**
The modules
couple using
their adapters.

ISS

Arm

Module

① **Robot**
The robotic
arm couples
the module.

Kibo (Experimental
Japanese Module)

ELM (Experimental
logistic module)

Unity node

**SEPTEMBER 2006**

**P3/P4 Truss and Solar Arrays**
The second and third port
truss segment was added, and
its solar panels were unfolded.
In June 2007, the second and
third starboard trus segment
(S3/S4) was docked, and its
solar panels were unfolded.

**FEBRUARY 2010**

**Tranquility Node**
Tranquility is a pressurized
module that supports many
of the space station's vital
systems. Attached to the node,
a cupola (right image) controls
the robotics. In February 2011,
Raffaello MPLM was used to
bring supplies to the ISS and
return the waste to Earth.

**APRIL 2016**

**BEAM**
The Bigelow Expandable
Activity Module (BEAM) is the
latest module docked with
the ISS. It is an expandable
experimental capsule that
inflates up to approximately
13 feet long and 10.5 feet in
diameter to provide a livable
volume for a crew member.

# SPYING ON THE UNIVERSE

Space telescopes such as the Hubble are artificial satellites put into orbit for observing different regions of the universe. Unlike telescopes on Earth, space telescopes are above the Earth's atmosphere. Therefore, they avoid the effects of atmospheric turbulence, which degrades the quality of telescopic images. Moreover, the atmosphere prevents the observation of stars and other objects in certain wavelengths (especially the infrared), which substantially decreases what might be seen in the heavens.

## HOW THE UNIVERSE WORKS

174

EXPLORATION MISSIONS

## ACCURATE CAMERAS

On the Hubble telescope, put into orbit on April 25, 1990, by NASA and ESA, the place of human observers is occupied by sensitive light detectors and cameras that photograph views of the cosmos. In 1993, due to a fault with its main mirror, corrective lenses (COSTAR) were installed to correct its focus.

14 m (46 ft)

4.26 m (14 ft)

### ENTRANCE
Opened during observations to allow light to enter.

---

### HOW IMAGES ARE CAPTURED
Hubble uses a system of mirrors that capture light and converge it until it becomes focused.

⟶ Direction of the light

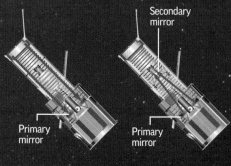

Secondary mirror

Secondary mirror

Primary mirror

Primary mirror

Primary mirror

WFPC

① **Entrance of light**
Light enters through the opening and reflects against the primary mirror.

② **Light ricochet**
The light then converges towards the secondary mirror, which returns it to the primary mirror.

③ **Image formed**
The rays of light concentrate on the focal plane, where the image is formed.

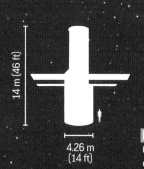

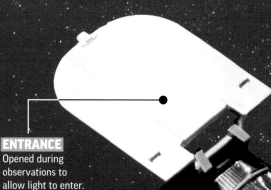

### OUTER COVER
Protects the telescope from the effects of outer space. During repair missions, the astronauts inspect it to look for particles and debris to be removed.

### SECONDARY MIRROR
After being reflected here, light reaches the camera.

## 11,000 KG
## (24,250 LB)
**HUBBLE'S WEIGHT ON EARTH.**

### WIDE FIELD PLANETARY CAMERA (WFPC)
Main electronic camera.

## HOW IMAGES ARE TRANSMITTED

**①  Hubble**
Instructions for the desired observation are uploaded to the telescope, which then transmits the image or other observational data after the observation is completed.

**②  TDRS satellite**
Receives the data from Hubble and sends them to a receiving antenna at the White Sands Test Facility in New Mexico.

**③  Earth**
From New Mexico, the information is transmitted to Goddard Space Flight Center in Greenbelt, Maryland, where it is analyzed.

## HUBBLE IMAGES

As they are outside the Earth's atmosphere, the clarity of images taken by Hubble is much better than those taken from telescopes on Earth. The Hubble can photograph a large variety of objects—from galaxies and clusters of galaxies to stars on the verge of exploding (such as Eta Carinae) and planetary nebulae (such as the Cat's Eye).

ETA CARINAE STAR     SUPERNOVA     CAT'S EYE NEBULA

## OTHER TELESCOPES

**CHANDRA**
Launched in 1999, it is the only X-ray observatory.

**SOHO**
Developed jointly by NASA and ESA, it allows scientists to view interactions between the Earth and the sun in detail. Placed into orbit in 1995.

**SPITZER**
Launched in August 2003, it observes the universe in infrared light.

**HIGH GAIN ANTENNA**
It receives orders from Earth and returns photographs as TV signals.

**SOLAR PANEL**
Power is provided by means of directional solar antennae.

**PRIMARY MIRROR**
Measuring 2.4 m (8 ft) in diameter, it captures and focuses light.

**COSTAR**
Optics device that corrected the original defective mirror fitted on Hubble.

Camera for blurry objects

# THE CHANDRA OBSERVATORY

In July 1999, the Chandra Observatory was put into orbit. This telescope can view the heavens using X-rays with an angular resolution of 0.5 arc-seconds, making it one thousand times more powerful than the first orbital X-ray telescope, named Einstein. This feature allows it to detect light sources that are 20 times more diffuse. The group tasked with constructing the X-ray telescope was responsible for developing technologies and processes that had never been applied.

## CUTTING-EDGE TECHNOLOGY

The satellite system provides the structure and equipment required for the telescope and scientific instruments to work as an observatory. To control the critical temperatures of its components, Chandra has a special system comprising radiators and thermostats. The satellite's electricity is supplied by solar panels and is stored in three batteries.

### HOW THE IMAGE IS CREATED

The information gathered by Chandra is extracted into images and tables with coordinates on the X- and Y-axes.

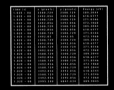

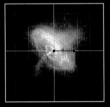

**① Table**
Contains the time, position and energy collected by Chandra's observations.

**② X-axis**
The data extends horizontally through the grid.

**③ Y-axis**
The data extends vertically through the grid.

## ① OBSERVATION

**The telescope's camera takes an X-ray image and sends it to the Deep Space Network for processing.**

Photographic camera

Solar panel

High-resolution mirror

X-rays

4 hierarchical hyperboloids

## ④ CHANDRA X-RAY CONTROL CENTER

**Tasked with ensuring the observatory's functionality and with receiving images. The operators are also responsible for preparing commands, determining the altitude and monitoring the condition and safety of the satellite.**

## ③ JET PROPULSION LABORATORY

**Information is received from the Deep Space Network and processed.**

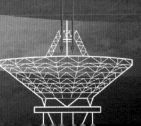

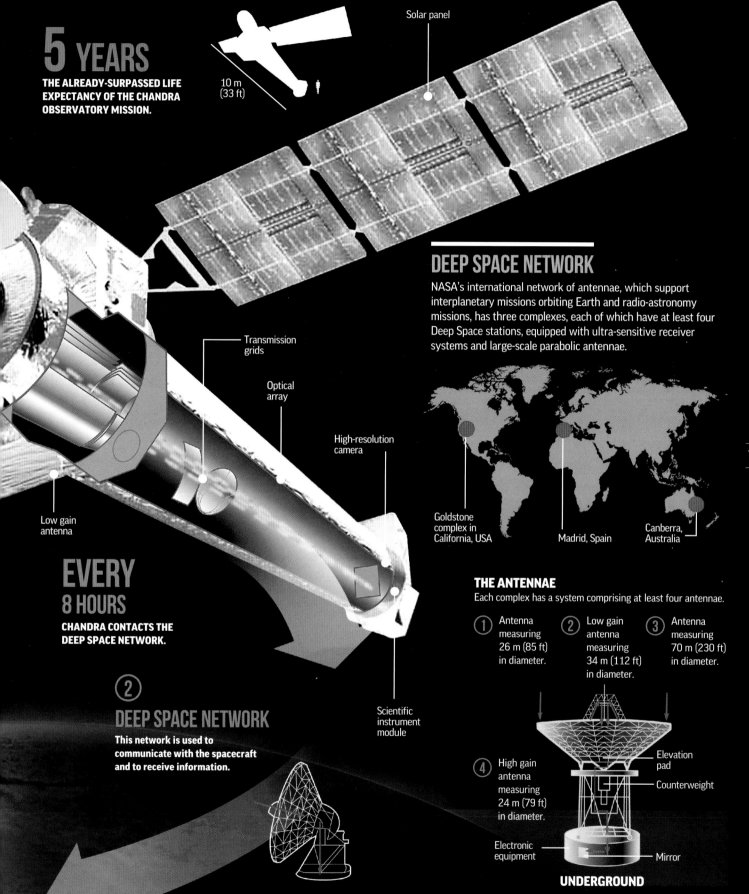

# 5 YEARS

**THE ALREADY-SURPASSED LIFE EXPECTANCY OF THE CHANDRA OBSERVATORY MISSION.**

10 m (33 ft)

Solar panel

Transmission grids

Optical array

High-resolution camera

Low gain antenna

# EVERY 8 HOURS

**CHANDRA CONTACTS THE DEEP SPACE NETWORK.**

② DEEP SPACE NETWORK

**This network is used to communicate with the spacecraft and to receive information.**

Scientific instrument module

## DEEP SPACE NETWORK

NASA's international network of antennae, which support interplanetary missions orbiting Earth and radio-astronomy missions, has three complexes, each of which have at least four Deep Space stations, equipped with ultra-sensitive receiver systems and large-scale parabolic antennae.

Goldstone complex in California, USA

Madrid, Spain

Canberra, Australia

### THE ANTENNAE
Each complex has a system comprising at least four antennae.

① Antenna measuring 26 m (85 ft) in diameter.

② Low gain antenna measuring 34 m (112 ft) in diameter.

③ Antenna measuring 70 m (230 ft) in diameter.

④ High gain antenna measuring 24 m (79 ft) in diameter.

Elevation pad

Counterweight

Electronic equipment

Mirror

**UNDERGROUND**

# VOYAGER PROBES

**The Voyager 1 and 2 space probes were launched by NASA to study the outer Solar System.** Launched in 1977, they reached Saturn in 1980 and Neptune in 1989, and they are currently continuing on their journey beyond the Solar System. Both probes have become the furthest reaching artificial instruments launched by mankind into space.

## PIONEER 10 AND 11
Pioneer 10 was the first spacecraft to fly over Jupiter, in 1973, and to study Saturn, in 1979. It was followed by Pioneer 11 in 1974, which lost communication in 1995.

## VOYAGER INTERSTELLAR MISSION

When Voyager 1 and 2 left the Solar System, the project was renamed the Voyager Interstellar Mission. Both probes continue studying the fields that they detect, in search of the heliopause – the boundary between the end of the Sun's reach and outer space.

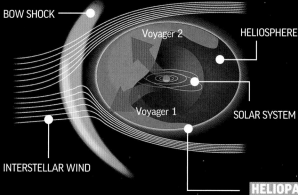

BOW SHOCK

Voyager 2

HELIOSPHERE

Voyager 1

SOLAR SYSTEM

INTERSTELLAR WIND

### HELIOPAUSE
Boundary between the area of the Sun's influence and outer space.

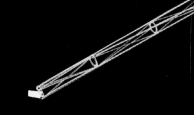

Earth
Jupiter
Saturn
Uranus
Neptune

Voyager 2

Voyager 1

### TRAJECTORY
The Voyager probe passed by Jupiter in 1979 and by Saturn in 1980. Voyager 2 did the same and arrived at Uranus in 1986, and Neptune in 1989. Both are still active.

## BEYOND THE SOLAR SYSTEM
Once outside the heliopause, Voyager can measure waves that escape the Sun's magnetic field, from the so-called bow shock, an area where solar winds suddenly decrease due to the disappearance of the Sun's magnetic field.

# 36 YEARS
**HAVE PASSED SINCE THE VOYAGER PROBES WERE SENT INTO SPACE.**

## LANDMARKS

### 1977
**Launches**
The Voyager 1 and 2 probes were launched by NASA from Cape Canaveral in Florida. This marked the beginning of a long, successful mission that is still ongoing.

### 1977
**Photo of the Earth and the Moon**
On September 5th, Voyager 1 sent photographs of the Earth and the Moon, demonstrating that it is fully functional.

### 1986
**Uranus encounter**
On January 24th, Voyager 2 reached Uranus. It sent photographs of the planet and measurements of its satellites, rings and magnetic fields back to Earth.

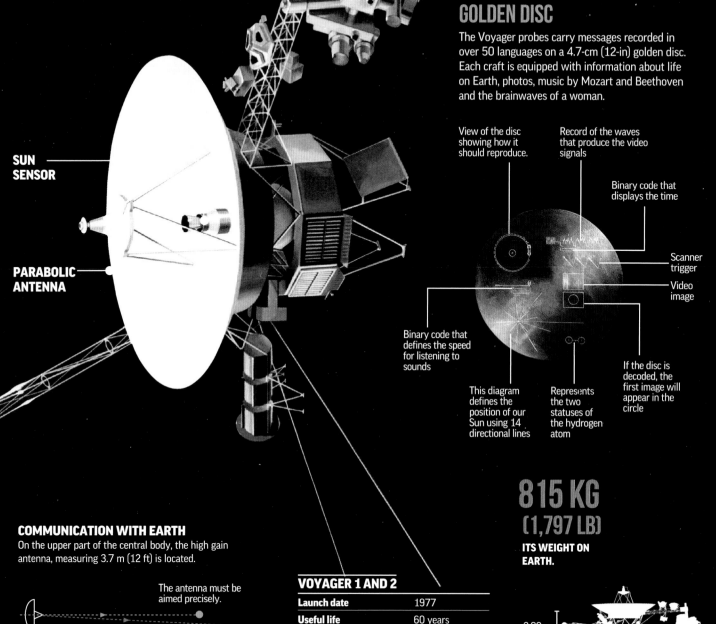

## GOLDEN DISC

The Voyager probes carry messages recorded in over 50 languages on a 4.7-cm (12-in) golden disc. Each craft is equipped with information about life on Earth, photos, music by Mozart and Beethoven and the brainwaves of a woman.

View of the disc showing how it should reproduce.

Record of the waves that produce the video signals

Binary code that displays the time

Scanner trigger

Video image

Binary code that defines the speed for listening to sounds

This diagram defines the position of our Sun using 14 directional lines

Represents the two statuses of the hydrogen atom

If the disc is decoded, the first image will appear in the circle

**SUN SENSOR**

**PARABOLIC ANTENNA**

## COMMUNICATION WITH EARTH

On the upper part of the central body, the high gain antenna, measuring 3.7 m (12 ft) is located.

The antenna must be aimed precisely.

**Antenna**
A sensor records the position of the Sun.

If the antenna deviates off position, the information does not reach its destination.

## VOYAGER 1 AND 2

| Launch date | 1977 |
|---|---|
| Useful life | 60 years |
| Weight | 815 kg (1,797 lb) |
| Power source | Plutonium |
| Organization | NASA |

# 815 KG
## (1,797 LB)
**ITS WEIGHT ON EARTH.**

2.89 m (9 ft)

3.7 m (12 ft)

---

**1987**
**Observation of a supernova**
Supernova 1987A appeared in the Large Magellanic Cloud. A high quality photograph was taken by Voyager 2.

**1989**
**Color photo of Neptune**
Voyager 2 is the first spacecraft to observe Neptune. It also photographed its largest moon, Triton, from close up.

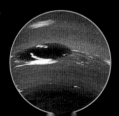

**1998**
**Beats the record set by Pioneer 10**
Pioneer 10 flew over Jupiter in 1973. On 17 February, Voyager 1 passes the planet and becomes the farthest reaching spacecraft in history.

# SETI PROJECTS

If intelligent life exists in the Universe, it is likely that it has attempted to communicate using radio signals. Since the 1960s, different SETI (Search for Extraterrestrial Intelligence) programs have focused on sweeping space with potent radio telescopes in an attempt to locate radio waves emanating from other civilizations. At present, the search has been unfruitful.

## 350 DISHES

THE NUMBER OF ANTENNAS THAT WILL COMPRISE ATA IN THE FUTURE. ITS SWEEPING AREA WILL BE EQUIVALENT TO ONE ANTENNA OF 114 METERS IN DIAMETER.

## THE ALLEN TELESCOPE ARRAY (ATA)

The Hat Creek Radio Observatory is home to the ATA, which consists of 42 radio telescopes similar to this one located in a remote valley 300 miles northeast of San Francisco, California.

## TIRELESS SEARCH

Up until 1985, the SETI Institute was comprised of different NASA-sponsored programs. Starting in 1985, SETI began operations focused solely on the search for extraterrestrial life forms in space. This private non-profit institute, headed by astronomer Frank Drake, has been financed for years by private donors, including Paul Allen, co-founder of Microsoft, who gifted $24 million towards the construction of the Allen Telescope Array. Along with the SETI Institute, other projects continue to sweep space for radio waves.

### BIG EAR

The first radio telescope used in monitoring space in search of intelligent life was inaugurated in 1963. It swept without success for 22 years before being decommissioned.

### SETI@HOME

A scientific experiment, based at UC Berkeley, that uses Internet-connected computers from around the world to process data collected from the Arecibo Observatory radio telescope.

### ATA

Thanks to the efforts of the SETI Institute and UC Berkeley, in 2007 the Allen Telescope Array—a potent radio telescope that combines the signal of multiple antennas—became operational.

### BREAKTHROUGH LISTEN

Since 2015, this new $100 million scientific research program, which aims at finding evidence of civilizations in the 100 galaxies closest to ours, utilizes two of the largest radio telescopes: Green Bank in West Virginia and Parkes in Australia.

## ARECIBO MESSAGE

On November 16, 1974, the strongest interstellar radio message carrying basic information about humanity and Earth was sent into space by the Arecibo radio telescope in Puerto Rico. Designed by Frank Drake, Carl Sagan and other astronomers, this pictogram consisted of 1,679 binary digits and was aimed at the current location of M13 some 25,000 light years away in the hope that it might be captured by an intelligent life form.

**Being human**
On the left, the average human height (1.764 meters/5.77 feet) and on the right, the global population at the time: 4,292,853,750.

### TRANSCRIPTION
The message included information on the Solar System, the Earth and the human species. The 1,679 bits or binary digits (the product of two prime numbers) were arranged from left to right and vertically and horizontally in 23 rows and 73 columns, depicting a design with different symbolic elements.

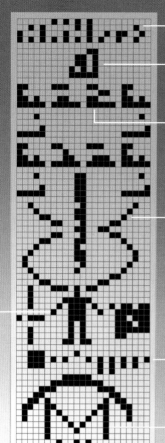

**Numbers**
1 to 10 in binary format.

**DNA molecules**

**Nucleotides**

**Double helix**
Of DNA.

**Solar System**
The Sun and the 9 planets (Pluto is included).

**Telescope**
Message sender.

## THE WOW! SIGNAL

No radio signal has been captured of a possible extraterrestrial transmission. The only signal received which can be considered an anomaly was captured on August 15, 1977 by the Big Ear radio telescope. The entire signal sequence lasted for 72 seconds coming from the Sagittarius constellation with intensity 30 times greater than the ordinary ambient noise of deep space. Professor Jerry R. Ehman of Ohio State University discovered the anomaly. After years of investigation, the signal has never been observed again and so remains unexplained.

### THE NAME
It's called the WOW! signal because of Ehman's comment written in red ink on the continuous paper that registered the

# CLOSER TO THE SUN

The space probe Ulysses was launched from the Space Shuttle on October 6th, 1990. It completed its first orbit around the Sun in 1997 and since then has carried out one of the most in-depth studies ever about our star. The probe's orbits allow it to study the heliosphere at all latitudes, from the equator to the poles, in both the northern and southern hemispheres of the Sun. The joint NASA and ESA mission is the first to orbit around the poles of the Sun. It orbits the Sun at 15.4 km/s (10 miles per second).

**SWOOPS**
An instrument that studies the ionic composition of the solar wind and the particle material.

**PASSES OVER THE SOLAR NORTH POLE**
June-October 1995
September-December 2001
November 2007-January 2008.

① Beginning of the first orbit: 1992

② Beginning of the second orbit: 1998

③ Beginning of the third orbit: 2004

**SUN**   **EARTH**

**JUPITER**
Flies by the planet and uses it for a gravity assist

**PASSES OVER THE SOLAR SOUTH POLE**
June-November 1994 / September 2000
January 2001 / November 2006-April 2007.

100 days

**HIGH-GAIN ANTENNA**
The antenna is used for communication with Earth stations.

## FIRST ORBIT
ORDER OF THE HELIOSPHERE
Ulysses completed its first solar orbit in December 1997 after having passed over the north pole. The heliosphere's structure was seen to be bimodal— that is, the solar winds were faster at greater inclinations of the orbit (beginning at 36°). During the first orbit, there was relatively little solar activity.

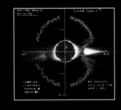

## SECOND ORBIT
HELIOSPHERE CHAOS
The information obtained by the Ulysses probe in the year 2000 showed a structural change in the solar wind during the period of maximum solar activity. Ulysses did not detect patterns in which wind speed corresponded with inclination, and in general the solar wind was slower and more variable.

## THIRD ORBIT
After having survived the difficult pounding of the solar activity during its second orbit, the Ulysses probe began a third orbit around the Sun's poles in February 2007. Solar activity was expected to be at a minimum, as it was in 1994, but the poles of the magnetic field are reversed.

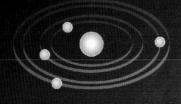

**THERMOELECTRIC RADIOISOTOPE GENERATOR**
provides electric energy for propelling the spacecraft in space.

## DUST

An internal device to study the energy composition of the heliosphere's particles and cosmic dust.

## RADIAL ANTENNA

contains four devices for different experiments.

## GRM

A device that studies the gamma rays emitted by the Sun.

## VHM

A device for studying the magnetic field of the heliosphere.

## 15.4 KM
### (10 MI) PER SECOND

**THE VELOCITY REACHED BY THE ULYSSES PROBE.**

## URAP

is used to measure the radio waves and plasma in the solar wind.

## HI-SCALE

Device designed to measure the energy present in ions and electrons of the interplanetary medium.

## REACTION TANK

A tank of fuel used for correcting the probe's orbit.

## ANTENNA CABLE CONTROL

A device onboard the spacecraft to change the position of the antennas.

## GOLD COVERING

It serves as insulation to help maintain the spacecraft's instruments at a temperature below 35° C (95° F) while the fuel is kept at a temperature above 5° C (41° F).

## ANTENNA CABLE

There is one on each side of the spacecraft. They are deployed after liftoff.

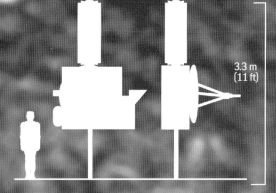

3.3 m
(11 ft)

## TECHNICAL SPECIFICATIONS:

| Launch date | October 6th, 1990 |
|---|---|
| Weight when launched | 370 kg (815 pounds) |
| Weight of the instruments | 550 kg (1,200 pounds) |
| Orbital inclination | 80.2° with respect to the ecliptic |
| Organization | NASA and ESA (joint mission) |

## THE RED PLANET

Among the planets in our Solar System, only the surface features of Mars are visible through a telescope. Thus, it became an object of study long before the beginning of the space age.

# TRAVELING TO MARS AND OTHER WORLDS

By its proximity and characteristics, Mars is the planet that arouses more interest, with even its colonization planned in the near future. Venus is somewhat closer to Earth, but its topography is hidden under a thick, opaque atmosphere. Jupiter and Saturn, much more distant, are surrounded by swirling clouds that hinder their observation. And Mercury is small and difficult to examine.

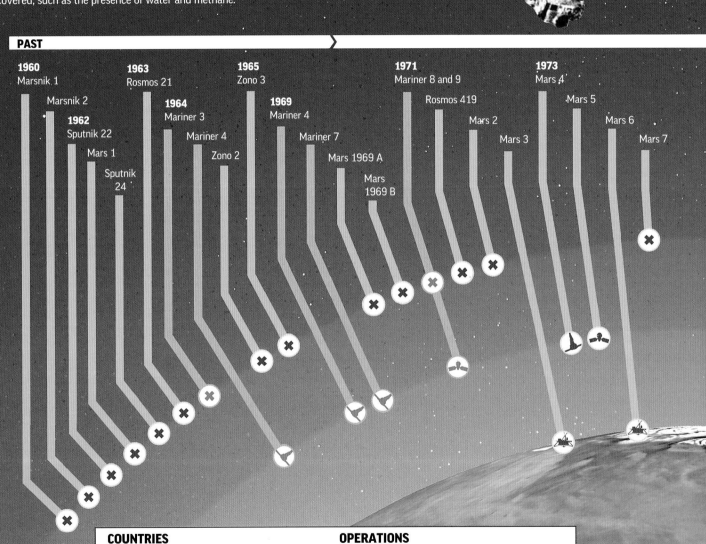

the planet, while others landed. In the future, robotic exploration will allow ships with humans onboard to be able to go there and return to Earth.

the first photos of the surface of Mars.

## PAST, PRESENT AND FUTURE

The ex Soviet Union, the United States, Japan, and the European Union have tried reaching Mars by different means. When orbiting the red planet and exploring it with autonomous vehicles, surprising aspects of the Martian environment were discovered, such as the presence of water and methane.

**PAST**

**1960** Marsnik 1

Marsnik 2

**1962** Sputnik 22

Mars 1

Sputnik 24

**1963** Rosmos 21

**1964** Mariner 3

Mariner 4

Zono 2

**1965** Zono 3

**1969** Mariner 4

Mariner 7

Mars 1969 A

Mars 1969 B

**1971** Mariner 8 and 9

Rosmos 419

Mars 2

Mars 3

**1973** Mars 4

Mars 5

Mars 6

Mars 7

**COUNTRIES**

- Soviet Union
- United States
- European Union
- Russia
- Japan

**OPERATIONS**

- Over flight
- Orbiter
- Lander
- Rover
- Failed

# A HUMID AND WARM PAST

The large amount of information transmitted by the probes and rovers has allowed us to decipher how the current Mars is and to imagine what this planet was like in the distant past. The images show evidence that liquid has flowed on its surface: valleys produced by the water course and typical formations of lakes. Also, minerals have been detected that need the existence of liquid water for their formation and the Mars Odyssey localized hydrogen in the first meters of soil, further proof that in the past there was liquid water on Mars. It is believed that this was possible about 3.5 billion years ago thanks to a greenhouse effect generated by volcanic activity and the impact of large meteorites.

## PATHFINDER
The first robot capable of moving over the Martian ground, landed on the planet in 1996.

## EXOMARS
The European robot will drill the ground and trace a map in 3D while it communicates with its mothership.

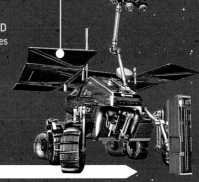

## PRESENT AND FUTURE

**2016**
**ExoMars**
It has the Trace Gas Orbiter (TGO) probe, the Schiaparelli landing module (which crashed in October 2016) and two rovers to be launched on Mars in 2020. The objective is to measure several gases (especially methane, a sign of life) under the ground and in the atmosphere.

**2018**
**InSight**
NASA will send a lander equipped with a seismometer to the Martian surface to analyze the interior of the planet.

**2020**
**Mars 2020 Rover**
This rover, similar to Curiosity, will have the mission to search for biomarkers, that is, present and past traces of the existence of life on Mars.

**2030**
**Manned Trip**
It will be made in an MPVC capsule (heir to Apollo).

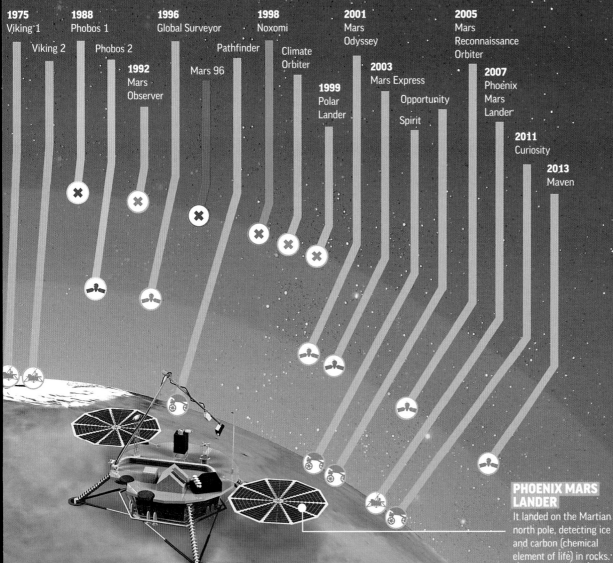

| 1975 | 1988 | | 1996 | | 1998 | | 2001 | | 2005 | |
| Viking 1 | Phobos 1 | | Global Surveyor | | Noxomi | | Mars Odyssey | | Mars Reconnaissance Orbiter | |

Viking 2    Phobos 2    Pathfinder    Climate Orbiter

**1992**
Mars Observer

Mars 96

**1999**
Polar Lander

**2003**
Mars Express

Opportunity

Spirit

**2007**
Phoenix Mars Lander

**2011**
Curiosity

**2013**
Maven

## PHOENIX MARS LANDER
It landed on the Martian north pole, detecting ice and carbon (chemical element of life) in rocks.

# MARS IN THE SIGHTS

**There was a time when it was thought that Mars, our closest neighbor, harbored life.** Perhaps for this reason it is the planet that has been most explored by various spacecraft from the decade of the 1960s onward, and it is therefore the one we know the best, apart from the Earth. Mariner IX in 1971 and Vikings I and II in 1976 revealed the existence of valleys and immense volcanic mountains. In 2001 the United States launched the Mars Odyssey mission, which indicated that liquid water exists at great depths.

## MARS ODYSSEY MISSION

Named after *2001: A Space Odyssey*, the probe was launched by NASA from Cape Canaveral on April 7, 2001. It entered into Martian orbit in October of the same year. The Mars Odyssey was designed for a number of functions, such as taking images in the visible and infrared spectrum, studying the chemical composition of the planet's surface, and investigating the existence of possible sources of heat. One of its purposes was also to find traces of hydrogen and thus water on Mars. Finally, the Mars Odyssey was used in support tasks for other Mars missions, acting as a radio-signal repeater between Earth and probes on the Martian surface.

**HINGE MECHANISM**

**GRS**
**Gamma-Ray Spectrometer**
weighs 30 kg (70 pounds) and consumes 30 watts. It measures the abundance and distribution of 20 chemical elements on Mars.

**THERMAL SHIELD**

**DOOR**

**SUPPORT**

**HEAD OF THE GAMMA-RAY SENSOR**

### LAUNCH
**APRIL 7, 2001**
The spacecraft Mars Odyssey is launched toward Mars atop a Delta 2 rocket.

### MARS ARRIVAL
**OCTOBER 24, 2001**
The spacecraft Mars Odyssey reaches the orbit of Mars and begins its scientific studies.

**MARS ODYSSEY**

**MARS AT TIME OF LAUNCH**

**SUN**

**EARTH AT TIME OF ARRIVAL**

**EARTH**

**MARS**

### CURRENT LOCATION OF THE ODYSSEY
It is orbiting Mars. It discovered the existence of ice, which was seen as a potential source of water for a future manned mission to the Red Planet.

### MAY 2001
The spacecraft tests its cameras by sending an image of the Earth at a distance of 3 million km (2 million mi).

### JUNE 2001
The gamma-ray spectrometer's protective hood is opened. The sensor begins to work.

### JULY 2001
The probe activates its auxiliary engines to adjust its trajectory. The thrust lasts 23 seconds.

### SEPTEMBER 2001
The probe begins to use the atmosphere to brake its speed, shape its orbit, and begin its mission.

## EARTH SEEN FROM MARS

Seen from Mars, the Earth is a magnificent blue star. From there, one can see the linked motions of the Earth and the Moon, as well as the combined phases of both. This photograph was taken by the Mars Odyssey in April 2006. Thanks to the spacecraft's infrared vision system, it was able to detect the temperatures on Earth, later confirmed by Earth-based sensors.

A view of Earth from Mars as recorded by Mars Odyssey.

### TECHNICAL SPECIFICATIONS:

| | |
|---|---|
| **Launch** | April 7, 2001 |
| **Arrived on Mars** | Oct. 24, 2001 |
| **Cost of the** | $332 million |
| **Weight** | 725 kg (1,600 pounds) |
| **Useful life** | 15-20 years |

2.20 m (7 ft)

2.60 m (8.5 ft)

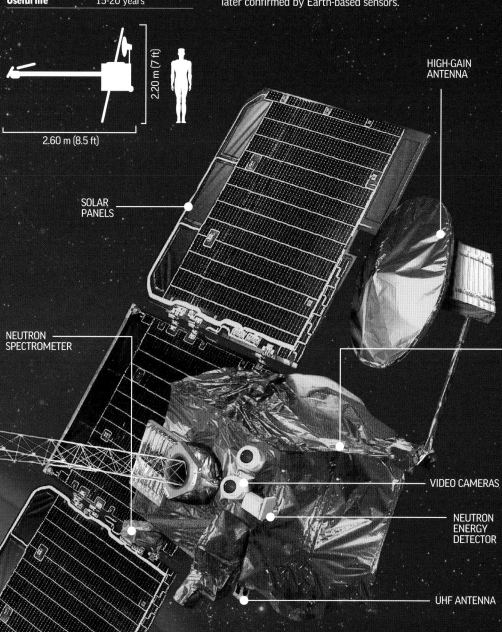

SOLAR PANELS

NEUTRON SPECTROMETER

HIGH-GAIN ANTENNA

VIDEO CAMERAS

NEUTRON ENERGY DETECTOR

UHF ANTENNA

## DISCOVERY

The new observations of Mars made by the Odyssey suggest that the north pole has about one third more underground ice than the south pole. Scientists also believe that microbial life could have developed on a planet other than Earth.

### MARIE
**An experiment measuring Mars' radiation environment**
It weighs 3 kg (7 pounds) and consumes 7 watts. It is supposed to measure radiation produced by the Sun or other stars and celestial bodies that reach the orbit of Mars.

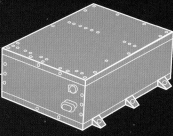

# 7 MONTHS
**THE TIME IT TOOK MARS ODYSSEY TO REACH ITS TARGET.**

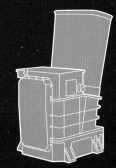

### THEMIS
**Thermal Emission Imaging System**
Weighing 911 kg (2,000 pounds) and consuming 14 watts, this camera operates in the infrared spectrum. Its images allow conclusions to be drawn about the composition of the surface based on the spectrum of the infrared image and on the recorded temperature.

Unlike the Earth, basalt dunes are common on Mars. The surface is flat and reminiscent of a desert.

# MARS RECONNAISSANCE ORBITER

**Since the first spacecraft, such as Mariner in the 1960s, the contribution to science made by space probes has been considerable.** Mostly solar-powered, these unmanned machines are equipped with sophisticated instruments that make it possible to study planets, moons, comets and asteroids in detail. One particularly renowned probe is the Mars Reconnaissance Orbiter (MRO) launched to study Mars from close up in 2005.

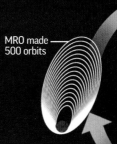

**APPROACH TO MARS**

MRO made 500 orbits

Ⓒ **Final orbit**
It traveled along an almost circular orbit, suitable for obtaining data.

Orbit

Ⓑ **Braking**
To get closer to the planet, the spacecraft slowed down over a six-month period.

Mars

Ⓐ **Start**
The probe's first orbit traveled along an enormous elliptical path.

## MARS RECONNAISSANCE ORBITER

The main objective of this orbiting probe is to seek out traces of water on the surface of Mars. The probe was launched in summer 2005 by NASA and reached Mars on March 10, 2006. It traveled 116 million km (72 million mi) in seven months. After more than ten years, the mission continues.

# 116 MILLION KM
## (72 MILLION MI)

**TRAVELLED BY THE PROBE ON ITS JOURNEY TO MARS.**

Mars' orbit

SUN

Earth

Earth's orbit

Mars

Ⓢ **Scientific phase**
The probe began its analysis phase on the surface of Mars. It found evidence of water.

④ **Arrival on Mars**
In March 2006, MRO passed into the southern hemisphere of Mars. The probe slowed down considerably.

① **Launch**
Took place on August 12, 2005 from Cape Canaveral, USA.

② **Cruising**
The probe travelled for seven and a half months before reaching Mars.

③ **Path correction**
Four maneuvers were made to ensure the correct orbit was reached.

# 1,031 KG
## (2,273 LB)

**WEIGHT ON EARTH.**

### TECHNICAL DATA:

| | |
|---|---|
| **Weight with fuel** | 2,180 kg (4,806 lb) |
| **Panel resistance** | Up to -200° C (-328° F) |
| **Launch rocket** | Atlas V-401 |
| **Duration of the mission** | 2006 - Ongoing |
| **Cost** | US$ 720 million |

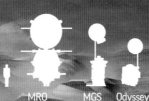

MRO     MGS     Odyssey

# ON MARS

The main objective of the MRO was to find evidence of water on the surface of Mars. In doing so, the evolution of the planet could be explained. The probe's devices facilitate high-resolution imagery of the surface and analysis of minerals. It also created daily climatic maps of Mars.

## 3,744
### CELLS ON EACH PANEL CONVERT SOLAR ENERGY INTO ELECTRICITY.

### HIGH GAIN PARABOLIC ANTENNA
Its data transfer capacity is 10 times greater than the capacity of previous orbiters.

SOLAR PANEL

## SOLAR PANELS
The probe's main power source is the Sun. The craft has two solar panels with a total surface area of 40 m² (430 sq ft).

### Opening the panels
The panels are opened while in orbit.

They also move from left to right.

Once unfolded, they use an axis.

They begin to unfold upwards.

The panels are almost closed.

RADAR SHARAD

SOLAR PANEL

## INSTRUMENTS
Used at the same time, HiRISE, CTX and CRISM offer very high quality information of a given area.

### HIRISE HIGH-RESOLUTION CAMERA
Provides details on geological structures and has considerably improved resolution when compared to previous missions.

**HiRISE**
Mars Reconnaissance Orbiter (2005)

**MGS**
Mars Global Surveyor (1996)

### MGS
Observes Mars' atmosphere.

### MARCI
Provides the images with color

### CRISM SPECTROMETER
Divides visible and infrared light in the images into various colors that identify different minerals.

### CTX CONTEXT CAMERA
Offers panoramic views that help to provide context to the images captured by HiRISE and CRISM.

The type of image taken by the CTX that helps to provide context to an image taken by HiRISE.

Detailed image taken by HiRISE.

HiRISE          CRISM          CTX

30 cm (12 in)/pixel          150 cm (60 in)/pixel

# MARTIAN ROVERS

**Spirit and Opportunity, the twin robots launched in June 2003 from Earth, arrived on Mars in January 2004.** They were the first to travel on the surface of the red planet. Both form part of NASA's Mars Exploration Rovers mission and are equipped with tools to drill rocks and collect samples from the ground for analysis.

## WATER AND LIFE ON MARS

The main objective of the mission was to find evidence of past water activity on Mars. Although the robots have found evidence of this, they have been unable to find living microorganisms, given that the ultraviolet radiation and oxidizing nature of the soil make life on Mars impossible. The question that remains unanswered is whether life may have existed on Mars at some stage in the past. And what's more, whether life currently exists in the subsoil on Mars, where conditions may be more favorable.

1.5 m (5 ft)

## 155 KG
(342 LB)

**WEIGHT ON EARTH.**

### TECHNICAL SPECIFICATIONS:

| Landing date | Spirit: January 3, 2004 |
| | Opportunity: January 24, 2004 |
| Cost of the mission | US $ 820 million |
| Progress per day | 100 m (328 ft) |
| Plutonium load | Each spacecraft carries a 2.8 g (0.1 oz) load |
| Duration of the mission | Spirit: communication lost in 2010 |
| | Opportunity: operational |

## HOW TO REACH MARS

The journey to Mars took seven months. Once inside Mars' atmosphere, a parachute is deployed to slow down the descent.

Aeroshell

Parachute

**① Deceleration**
130 km (81 mi) from the surface, the aeroshell slows down from 16,000 to 1,600 km/h (9,941 to 994 mph).

**② Parachute**
10 km (6.2 mi) from the surface, the parachute opens to slow down the descent.

**③ Descent**
The shield that offered the rover heat protection separates from the input module.

Input module

**④ Rockets**
10-15 m (33-49 ft) from the surface, two rockets are ignited to slow down the descent. Two airbags are then inflated to surround and protect the landing gear.

**⑤ Airbags**
The landing gear and airbags detach from the parachute and fall to Mars' surface.

Descent rockets

**⑥ Landing**
The airbags deflate. The 'petals' that protect the spacecraft open. The vehicle emerges.

Vectran airbags

**⑦ Instruments**
The robot opens its solar panels, the mast camera and its antennae.

Photograph of the surface taken by Spirit.

## 70,000

**IMAGES OBTAINED BY SPIRIT DURING ITS FIRST TWO YEARS.**

Footprint and photograph taken by Opportunity.

## 80,000

**IMAGES OBTAINED BY OPPORTUNITY DURING ITS FIRST TWO YEARS.**

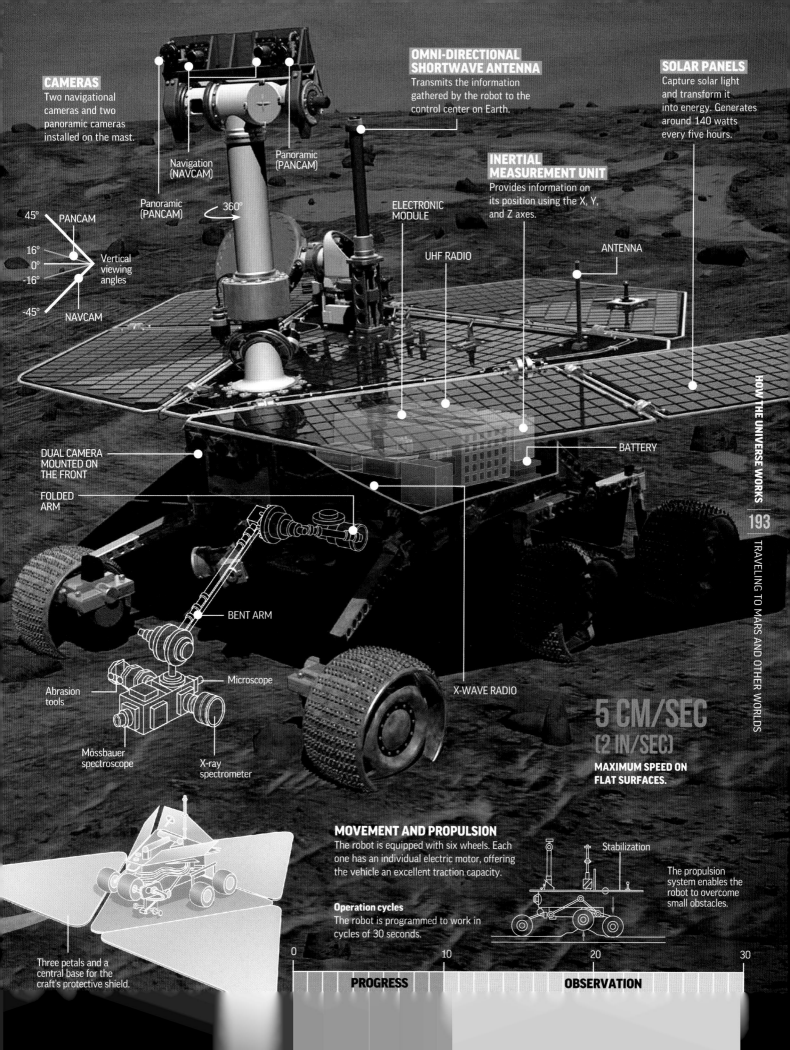

## CAMERAS
Two navigational cameras and two panoramic cameras installed on the mast.

Navigation (NAVCAM)

Panoramic (PANCAM)

Panoramic (PANCAM)

360°

45° — PANCAM
16°
0° — Vertical viewing angles
-16°
-45° — NAVCAM

## OMNI-DIRECTIONAL SHORTWAVE ANTENNA
Transmits the information gathered by the robot to the control center on Earth.

## SOLAR PANELS
Capture solar light and transform it into energy. Generates around 140 watts every five hours.

ELECTRONIC MODULE

UHF RADIO

## INERTIAL MEASUREMENT UNIT
Provides information on its position using the X, Y, and Z axes.

ANTENNA

BATTERY

DUAL CAMERA MOUNTED ON THE FRONT

FOLDED ARM

BENT ARM

Microscope

Abrasion tools

Mössbauer spectroscope

X-ray spectrometer

X-WAVE RADIO

## 5 CM/SEC
(2 IN/SEC)

**MAXIMUM SPEED ON FLAT SURFACES.**

Three petals and a central base for the craft's protective shield.

## MOVEMENT AND PROPULSION
The robot is equipped with six wheels. Each one has an individual electric motor, offering the vehicle an excellent traction capacity.

### Operation cycles
The robot is programmed to work in cycles of 30 seconds.

Stabilization

The propulsion system enables the robot to overcome small obstacles.

0       10       20       30

**PROGRESS**            **OBSERVATION**

# CURIOSITY

**The last NASA robot that landed on the red planet, Curiosity, arrived on Mars on August 5, 2012, and landed in the Galle crater.** The selection of the landing site was made by consensus among the scientific community, in view of the objectives proposed for the mission (Mars Science Laboratory), among which was the study of the habitability of the planet.

## IN SEARCH OF EVIDENCE

Since its landing, Curiosity has been moving towards the base of Mount Sharp and has been finding evidence that water flowed through the crater in the past. Images have been taken of a conglomerate of rocks similar to those found on the shore of a river on Earth. The shape and position of the rocks indicate that it was a shallow channel in which water flowed smoothly. The walls of the crater also tell us that the water flowed down the slope and accumulated just near the area where Curiosity landed.

### CHEMCAM
A laser located on the mast pulverizes some rocks that are then photographed and chemically analyzed.

**GALLE CRATER**

○ Landing zone

• Research zones

### LANDING ZONE
Galle crater, with a diameter of 150 km (93,2 mi), has a mountain in the center formed by mysterious rocky layers.

### MASTCAM
Main camera (it includes two digital cameras with HD video).

### REMS
Environmental sensors.

### MOVEMENT
It has an engine on each of its 6 wheels and moves 2 cm/sec.

### PROPULSION ENERGY
Plutonium 238 converted into electricity.

## 900 KG
### (1,984 LB)
**CURIOSITY WEIGHT.**

# OTHER FINDINGS

In addition to finding evidence of water and confirming that Mars is a place capable of life if some of its conditions were modified, the Curiosity robot made other important inquiries. For example, it verified that the high level of radiation would endanger the life of astronauts on the red planet. It also detected the presence of methane (in fluctuating intensity), a gas that on Earth usually has a biological origin.

## ENTERING THE ATMOSPHERE

**Parachute deployment**

**+0 sec**
**Speed:** 5.000 m/s
**Altitude:** 125 km

**+240 sec**
**Speed:** 470 m/s
**Altitude:** 10 km

**Thermal shield separation:** radar data collection.

**Capsule separation**

**+380 sec**
**Speed:** 0.75 m/s
**Altitude:** 20 m

## DESCENT SYSTEM

For the first time, an exploration robot has landed with the aid of a crane and cables, without the need of complex protection airbags. This assured a softer landing for Curiosity.

**Descent with rockets**

**Descent with the crane**

**Rockets are released**

**ROBOTIC ARM**
It has a freedom of movement of 5 degrees and reaches 2.2 m. It is used to get the instruments close to the ground and pick up rock samples for their analysis.

**MAHLI**
Large observation lens that can take color images with a high level of detail. Two white-light LED lamps allow for nighttime shots.

**APXS**
X-ray spectrometer: analyses of minerals and traces of elements.

## TECHNICAL SPECIFICATIONS:

**Landing date**
November 26, 2011

**Arrival to Mars**
August 6, 2012

**Autonomy**
1 Martian year (687 earth days or 23 months)

900 kg

3 m

# IS IT POSSIBLE TO COLONIZE MARS?

**If today one of the main scientific obsessions is the search for life on Mars, then its main objective for the future is to send manned missions including colonizing the red planet.** Mars' atmosphere, low temperatures and high levels of radiation, as well as the technical difficulties of the voyage, make for a very complicated challenge.

## SURVIVAL HANDICAPS

One of the greatest challenges facing the scientific community in colonizing Mars is the supply of water. Low temperature, present atmospheric pressure which is less than 1% of that of Earth and high levels of radiation prohibit liquid water on the planet's surface. Mars does not have a magnetic field and its atmosphere is not sufficient to protect the planet. Life on Mars for its early settlers will be confined to protective habitable modules until scientists can alter the long-term atmospheric conditions, which will make conditions habitable for humans.

**ASTEROIDS**
Contain large quantities of water. NASA is researching the possibilities of harvesting water from asteroids.

### ① WATER
We would need to search by perforating the subterranean surface, melting the ice from the surface of Mars or purifying residual waters using evapotranspiration.

### ② TERRAFORMING
The long-term objective would be to emit carbon dioxide or provoke the melting of the polar ice caps to create a greenhouse effect and "construct" a protective dense atmosphere similar to Earth.

**MODULE**
NASA is now adding focus on deep space habitat "modules" with recyclable air and water systems similar to those used in the International Space Station, without cargo supply deliveries from Earth.

## 6 MB
**CURRENT ATMOSPHERIC PRESSURE OF MARS.**

## WHICH CLOTHING WILL BE NECESSARY?

As long as there is no atmospheric pressure or oxygen, human visitors will have to continue using space suits. In 2011 NASA tried new pressurized designed in Antarctica. The NDX-1 prototype, created by aero spatial engineer Pablo de Leon at the University of North Dakota, United States, cost 100.000 dollars and was made with more than 350 materials, including carbon fibers and Kevlar (synthetic polyamide).

# -63° C
## (-81,4° F)
**AVERAGE EQUATORIAL
TEMPERATURE OF MARS.**

## A COMPLICATED TRIP

The first obstacle to overcome to land on Mars is the complicated voyage, which with today's technology can take up to seven months. Since the distance between Mars and Earth varies based on their orbits relative to each other (less than 60,000,000 km), an ideal time to depart is important. In addition to the physical challenges of weightlessness and a restricted diet facing astronauts on the long journey, there is also the greater risk of exposure to high levels of radiation.

### SUBTERRANEAN WATER
Finding sources of water is basic. Traces of salty water have been detected during the night at shallow depths, but during the day it dries up.

### ORION - MPCV
NASA in collaboration with The European Space Agency (ESA) have chosen a version of this spacecraft in order to transport astronauts to Mars' orbit, and eventually to the surface of the planet, by 2030.

### VASIMR
The Variable Specific Impulse Magnetoplasma Rocket (VASIMR) engine is a new type of electric thruster capable of converting gas such as hydrogen or ice into magnetized plasma. Coupled to a spacecraft, it can reduce the journey to 39 days.

# A FILM CHALLENGE

**NASA as well as several private initiatives have all started the race to send astronauts to Mars even though the enormous economic costs of realizing such a mission have slowed many of the projects.** The project will require not only a space vehicle capable of sending astronauts to such vast distances but also one that transports tons of materials and supplies, while also designed to be habitable once they have landed on the red planet.

## NASA: 2030 MISSION

According to projections by NASA, it is anticipated that the first manned missions to Mars will start in 2030. The first missions will be designed to orbit Mars and subsequent missions to land on the planet's surface. These missions will utilize NASA's recently created Space Launch System (SLS) rocket and a special version of the Orion spacecraft.

### EXOMARS

In the next couple of years, this joint endeavor between the European Space Agency (ESA) and the Russian Space Agency (Roscomos) will study in depth the geophysics and geochemistry of Mars. The data from these missions could be vital for future manned missions.

### GREENHOUSE

Due to the difficulties transporting supplies from Earth, any plan for habitation of Mars must include the cultivation of food in a greenhouse.

## MARS COLONIAL TRANSPORTER

One of the most ambitious private initiatives focused on colonizing the red planet by 2022 is led by South African businessman Elon Musk, the founder, CEO and lead designer at Space Exploration Technologies (SpaceX) as well as co-founder of PayPal, Tesla Motors and SolarCity, among others. Musk's plans on sending to Mars a great colony of up to 1 million inhabitants aboard spacecrafts that can transport up to 100 passengers. These spacecraft will utilize reusable propulsion rockets capable of completing several trips back to Earth for supplies.

## MARS ONE, A CONTROVERSIAL PROJECT

Lead by Dutch entrepreneur Bas Landsdorp, the Mars One project has been well covered by the media. Its objective is to create a human settlement on Mars in the coming decades by sending groups of four astronauts on one-way missions. The project was kicked off with a complex selection process to elect the 24 candidates who will travel to Mars even though the project faces much financial doubt. In fact, the project has experienced many delays to its manned missions and it is unclear which spacecraft and rocket system it will utilize.

### ROADMAP

**2011**
**Mars One founded**
Lansdorp plans and designs the mission.

**2013**
**Start Crew Selection**
More than 2,000 candidates are presented.

**2017**
**Start of Crew Training**
6 teams of 4 astronauts participate.

**2022**
**Demo Mission**
A demonstration mission will be sent to the surface of Mars.

**2024**
**ComSat Mission**
A satellite will be launched and placed into Mars orbit.

**2026**
**Rover & ComSat Mission**
To find teh best location for the settlement.

**2029**
**Cargo Missions**
Containing a second rover and other support units.

**2030**
**Outpost Operational**
The rover prepares the outpost.

**2031**
**Departure Crew One**
The Mars Transit Vehicle (MTV) is launched.

**2032**
**Landing Crew One**
The rover will bring the crew to the outpost.

**2033**
**Departure Crew Two**
And the cargo modules for the third crew.

### LIQUID WATER
Its presence on the surface of Mars is only feasible at temperatures at 4° C and a pressure of 500 millibars. Therefore, the atmosphere of the planet must be altered. Perhaps a human colony can accomplish this in the future.

## INSPIRATION MARS

In 2013, US multimillionaire Dennis Tito created a foundation focused on sending a manned mission to Mars by 2018. The ambitious plan is to send a man and woman – probably a married couple – on a round trip to Mars when planetary alignment is favorable in January 2018. Tito intends to take advantage of the alignment of heavenly bodies to fly around Mars and return to Earth in the relatively short time of 501 days. Lack of support from NASA has slowed the momentum of the project.

# JUPITER IN FOCUS

**The fifth planet of the Solar System was visited by Pioneer X and XI, Voyager 1 and 2, and Cassini.** However, the most significant visitor was Galileo, launched by NASA on October 18, 1989. Galileo consisted of an orbiter and an atmospheric probe. After a long voyage, the atmospheric probe penetrated some 200 km (125 mi) into the atmosphere of Jupiter on December 7, 1995, transmitting data about the atmosphere's chemical composition and Jupiter's meteorological activity. The orbiter continued sending information until it crashed into the gaseous giant on September 21, 2003.

## TRAJECTORY

Galileo was designed to study the atmosphere of Jupiter, its satellites, and the magnetosphere of the planet. To get there, it did not use a direct path but had to perform an assisted trajectory, passing by Venus on February. 10, 1990. Then it flew by the Earth twice and arrived at Jupiter on December 7, 1995. The probe succeeded in sending information of unprecedented quality with a low-gain antenna about the satellites of Jupiter, its moon Europa, and various examples of volcanic activity on its moon, named Io. It also contributed to the discovery of 21 new satellites around Jupiter. The mission was deactivated in 2003, and the vehicle was sent to crash into the planet. The purpose of this termination was to avoid future collision with its moon Europa that might have contaminated its ice; scientists believe that extraterrestrial microscopic life may have evolved on Europa.

**ATMOSPHERIC PROBE**
Released when Galileo arrived at Jupiter. It was used to study the planet's atmosphere.

LOW-GAIN ANTENNA

SO
PA

BOOSTERS

LOW
ANTE

MAGNETIC
SENSORS

**LAUNCH**
October 18, 1989
Galileo was launched by NASA from the space shuttle Atlantis with Jupiter as its destination.

**EARTH FLYBYS**
December 1990/August 1992
Galileo passes by the Earth on two occasions to get the necessary boost toward Jupiter.

**ARRIVAL AT JUPITER**
December 7, 1995
Galileo arrived at Jupiter and began the scientific studies that continued until 2003. It completed 35 orbits around the planet.

**VENUS FLYBY**
February 10, 1990
Galileo transmitted data from Venus.

**IDA FLYBY**
August 28, 1993
Galileo came close to the asteroid Ida.

**GASPRA FLYBY**
October 29, 1991
Galileo approached the asteroid 951 Gaspra.

**14 YEARS**
THE DURATION OF THE GALILEO MISSION—FROM OCTOBER 1989 TO SEPTEMBER 2003.

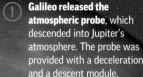

## DESCENT TO JUPITER

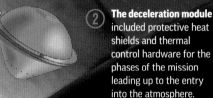

① **Galileo released the atmospheric probe**, which descended into Jupiter's atmosphere. The probe was provided with a deceleration and a descent module.

② **The deceleration module** included protective heat shields and thermal control hardware for the phases of the mission leading up to the entry into the atmosphere.

# GALILEO

In spite of its mission being plagued by technical problems, Galileo provided astronomers with a huge amount of information during its 35 orbits around Jupiter. The useful life of the probe extended five years longer than planned. The probe contributed to the discovery of 21 new satellites around Jupiter. Galileo sent large amounts of data and 14,000 images to Earth. It found traces of salt water on the surface of the moon Europa and evidence that it probably also exists on the moons Ganymede and Callisto. Likewise, it provided information about volcanic activity on the moon Io. It also showed an almost invisible ring around Jupiter consisting of meteorite dust. From the moment it was launched until its disintegration, the spacecraft traveled almost 4.6 billion km (2.9 billion mi) with barely 925 kg (2,000 pounds) of combustible fuel.

DECELERATION MODULE

ANTENNA

PARACHUTES

DESCENT MODULE

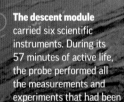

## TECHNICAL SPECIFICATIONS:

| Date of arrival | December 7, 1995 |
|---|---|
| Cost of the mission | US $1.5 billion |
| Useful life | 14 years |
| Weight without the probe | 2,223 kg (4,900 pounds) |
| Organization | NASA |

③ **A parachute 2.5 m (8 ft)** in diameter was used to separate the descent module from the deceleration module and to control the velocity of the fall during the atmospheric descent phase.

PARACHUTES

7 m (23 ft)

6.2 m (20 ft)

④ **The descent module** carried six scientific instruments. During its 57 minutes of active life, the probe performed all the measurements and experiments that had been planned by the scientists.

### ATMOSPHERE OF JUPITER
Composed of 90 percent hydrogen and 10 percent helium. The colors of the atmospheric clouds depend on their chemical composition. The clouds spread with the violent turbulence of the atmospheric winds.

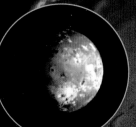

### IO
One of the moons of Jupiter. It is notable for its brilliant color, which is caused by various sulfur compounds on its surface. Io is 671,000 km (417,000 mi) from Jupiter and was discovered by Galileo Galilei in 1610.

# ATMOSPHERIC PROBE

Once Galileo arrived at the planet Jupiter, it released a small probe that fell through the atmosphere. This descent probe carried scientific instruments and the subsystems required to keep them active and transmit the data to the orbiter for storage for later transmittal to Earth. During its 57 minutes of active life in the Jovian atmosphere, the descent provided a number of discoveries, including a surprising lack of water in the upper layers of the Jovian clouds.

## TECHNICAL SPECIFICATIONS:

| Entry into the atmosphere | December 7, 1995 |
|---|---|
| Active life | 57 minutes |
| Weight | 339 kg (750 pounds) |
| Organization | NASA |

0.86 m (3 ft)

1.25 m (4 ft)

# THE MOON EUROPA

**Together with Mars, Europa is one of the main candidates to shelter life within the Solar System and that makes it one of the great focal points of interest in astronomy.** This satellite of Jupiter hides a large salty ocean under its icy surface where conditions could be conducive to the existence of some type of microorganism. Europa has already been observed by different probes in the last decades (Pioneer 10 and 11, the two Voyagers, Galileo, and New Horizons) and will soon be analyzed by two new missions from NASA and ESA.

## EUROPA ESSENTIAL DATA

**Diameter at the equator**
3,126 km (1,942 mi)

**Orbital speed**
13.74 km/sec (8.53 mi/sec)

**Escape velocity**
2,025 km/sec (1,258 mi/sec)

**Mass (Earth = 1)**
0.008

**Volume**
1,593 x 1010 m$^3$

**Gravity**
1,314 m/s$^2$ (4,311 ft/s$^2$)

**Density**
3.013 g/cm$^3$

**Temperature**
-223/-163° C (-369/-261° F)

## SLAVE OF JUPITER

Discovered by Galileo in 1610, it is believed that the Europa satellite is composed of a core of iron and nickel, a rocky mantle and an outer crust of ice between 10 and 30 km (6.21-18.64 mi) thick, under which there is a large salty ocean of approximately 90 km (55.92 mi). That water remains in liquid state thanks to the heat generated by the tides produced by the enormous gravitational force that Jupiter exerts and that even deforms the satellite. Some studies argue that this water has a high concentration of oxygen and could even harbor complex life forms.

### ATMOSPHERE OF EUROPA

It is very tenuous and is composed of oxygen of non-biological origin. Sunlight and space particles produce water vapor by striking the icy surface. Then hydrogen escapes into space and oxygen molecules form a weak atmosphere.

RAM POINTED
INSTRUMENTS

NADIR POINTED
INSTRUMENTS

### REASON HF ANTENNA

The probe has two ice-penetrating antennas (16 m) attached to the solar panels.

# 3.55 DAYS
**THE ORBITAL PERIOD OF EUROPA.**

## EUROPA CLIPPER AND LANDER

After ten years of redesignation due to strong readjustments in the mission budget, NASA will finally launch the Europa Clipper spacecraft in 2022, as long as the SLS rocket is available by that date. Although it will be in the orbit of Jupiter, this probe will periodically fly over the surface of Europa to analyze its frozen crust and collect data about its interior ocean. It will also serve to find the best location for the Europa Lander, which will be sent in 2025 with the mission of finding microorganisms.

# JUICE: THE ESA'S MISSION

The Jupiter Icy Moons Explorer is the mission that the European Space Agency (ESA) prepares to study the gas giant planet and its main satellites, including Europa. The JUICE probe will be launched in 2022 by the Ariane 5 rocket and is expected to reach Jupiter's orbit in 2030. The spacecraft includes cameras, spectrometers and an ice-penetrating radar intended to study the ocean that hides under the crust of the moon Europa.

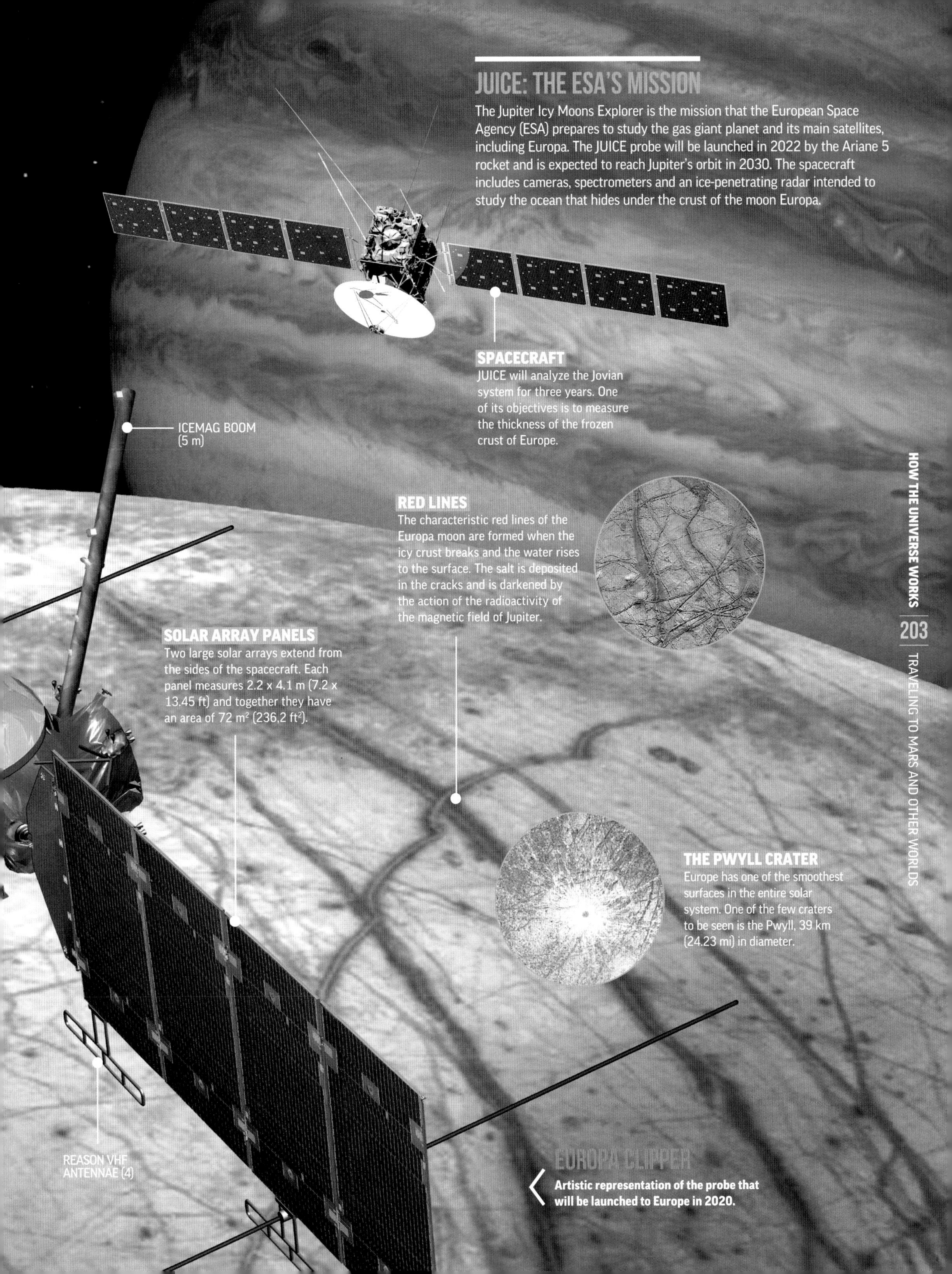

## SPACECRAFT
JUICE will analyze the Jovian system for three years. One of its objectives is to measure the thickness of the frozen crust of Europe.

## RED LINES
The characteristic red lines of the Europa moon are formed when the icy crust breaks and the water rises to the surface. The salt is deposited in the cracks and is darkened by the action of the radioactivity of the magnetic field of Jupiter.

**ICEMAG BOOM**
(5 m)

## SOLAR ARRAY PANELS
Two large solar arrays extend from the sides of the spacecraft. Each panel measures 2.2 x 4.1 m (7.2 x 13.45 ft) and together they have an area of 72 m² (236,2 ft²).

## THE PWYLL CRATER
Europe has one of the smoothest surfaces in the entire solar system. One of the few craters to be seen is the Pwyll, 39 km (24.23 mi) in diameter.

**REASON VHF ANTENNAE (4)**

**EUROPA CLIPPER**

**Artistic representation of the probe that will be launched to Europe in 2020.**

# A VIEW OF SATURN

**The longed-for return to Saturn was the result of a scientific alliance between NASA and the European Space Agency (ESA).** On October 15, 1997, after a number of years of development, the fruit of this collaboration lifted off toward this enormous gas giant. The mission of Cassini, the mother ship, was the exploration of Saturn. It carried a smaller probe, Huygens, that was to land on Saturn's largest moon, Titan, and transmit images and sounds from the surface. The Huygens probe accomplished this prodigious feat, demonstrating once again the capacity of humans to respond to the challenge of frontiers.

**THE RINGS OF SATURN** are a conglomerate of ice particles and powdered rock orbiting the planet. The rings are 4.5 billion years old.

## TRAJECTORY

The trajectory of Cassini-Huygens was long and complicated, because it included strategic flybys of Venus (1998 and 1999), Earth (1999), and Jupiter (2000). Each one of these encounters was used to increase the craft's velocity and to send the spacecraft in the appropriate direction (a maneuver known as a gravity assist). Finally, and after almost seven years, traveling some 3.5 billion km (2,2 billion mi), the spacecraft arrived at its destination. It brought an end to the long wait since the last visit of a probe to Saturn—the 1981 flyby by Voyager II.

**VENUS 1**
**April 18, 1998**
Cassini flies by Venus at an altitude of 284 km (180 mi).

**THE EARTH**
**August 1999**
Cassini flies by the Earth at an altitude of 1,171 km (730 mi).

**SATURN**
**June 2004**
After seven years en route, Cassini arrives at Saturn and enters into an orbit around it.

EXTENSION FOR THE MAGNETOMETER

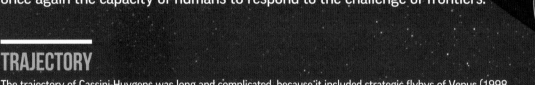

**VENUS 2**
**June, 1999**
Cassini flies by Venus at an altitude of 600 km (380 mi).

**JUPITER**
**December, 2000**
Cassini flies by Jupiter at an altitude of 9,723,896 km (6,042,000 mi).

### TRAJECTORY FOR SATURN AND TITAN
Here is a drawing showing some of the 74 orbits planned for the mission.

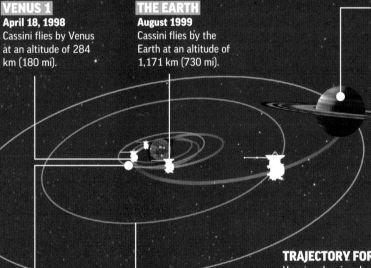

Meeting between Huygens and Titan

Equatorial Rotation

Upward Trajectory

**Titan's** Orbit

**Saturn** Seen from the North Pole

Initial Orbit

Occultation Orbit

Equatorial Rotation

Spacecraft Thruster (1 of 2)

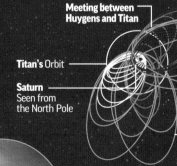

**PHOTO OF JUPITER AND IO**
The moon Io, the closest to the planet Jupiter, is composed of a rocky silicate material. The nucleus has a radius of 900 km (560 mi) and may consist of iron. This is the photo taken by the Cassini probe.

ANTENNA FOR THE RADIO SUBSYSTEMS AND THE PLASMA PROBES (1 OF 3)

# CASSINI-HUYGENS

The information sent by Huygens and relayed by Cassini took 67 minutes to travel from Saturn to the Earth. Although it could only see a small section of Titan, the apparatus was able to answer some key questions. For example, the probe did not find liquid, but it did find signs that the surface had a crust that was hard on top and soft underneath, which was flooded from time to time. Investigators said that Titan could have very infrequent precipitation, but when it occurred it could be abundant and cause flooding. Moreover, it appears that some of the conditions for life to arise exist on Titan, although it is too cold for life to have started.

# DESCENT ONTO TITAN

On January 14, 2005, the six instruments of Huygens worked without pause during the two-and-a-half-hour descent. They confirmed, for example, that the gaseous blanket that surrounds Titan consists primarily of nitrogen and that its yellowish color is caused by the presence of complex hydrocarbons, which are formed when sunlight breaks down atmospheric methane. The thermometer measured –203° C (–400° F) at an altitude of 50 km (31 mi), which was the lowest temperature recorded during the entire mission.

## TECHNICAL SPECIFICATIONS:

| Date of launch | October 15, 1997 |
|---|---|
| Begins Saturn orbit | July 1, 2004 |
| Closest approach | 5,600 kg (12,300 pounds) |
| Organizations | NASA and ESA |

6.7 m (22 ft)

4 m (13 ft)

HIGH-GAIN ANTENNA

**5,600 KG**
**(12,300 POUNDS)**
**WEIGHT ON EARTH.**

LOW-GAIN ANTENNA (1 OF 2)

RADAR

TELESCOPES

GTR (RADIOISOTOPE THERMOELECTRIC GENERATOR)

**350 KG**
**(770 PQUNDS)**
**WEIGHT ON EARTH.**

## TECHNICAL SPECIFICATIONS: HUYGENS

| Date of release | December 25, 2004 |
|---|---|
| Weight | 319 kg (703 pounds) |
| Organizations | NASA and ESA |
| Date of landing | Jan. 14, 2005 |
| Descent by parachute | 2.5 hours |

PLACEMENT OF HUYGENS ON CASSINI

22 feet (6.8 m)

9 feet (2.7 m)

### 1. Separation
The Huygens probe separates from Cassini.

### 2. Descent
lasted 150 minutes and came within 1,270 km (790 mi) of the surface.

### 3. First parachute
helped decelerate the probe during its fall.

### 4. Second parachute
replaced the first.

### 5. Third parachute
replaced the second.

### 6. Deploys its landing feet
The probe prepares for touchdown.

### 7. Impact on the surface
The spacecraft strikes the surface of Titan.

### 8. Landing
The probe took photographs and data from the surface of Titan.

**THE SURFACE OF TITAN** is obscured by a deep layer of clouds. It is possible that many chemical compounds similar to those that preceded life on Earth exist in a frozen state at high altitudes.

# TOWARD VENUS AND PLUTO

**The New Horizons mission, launched by NASA in January 2006, is a voyage that will carry the spacecraft to the limits of the Solar System and beyond.** The most important goal of the voyage is to visit Pluto, a dwarf planet (a designation made in 2006 by the International Astronomical Union). The ship flew past Jupiter to gain enough speed to get to Pluto in the year 2015. After a few months of observations of Pluto, which approached 12,450 km (7,736 mi), its voyage continued toward the region of the Solar System known as the Kuiper belt.

**SPECTROMETER 1**
will study the interaction of Pluto with the solar wind to determine if it possesses a magnetosphere.

## NEW HORIZONS MISSION

An unmanned space mission by NASA whose destination is to explore Pluto and the Kuiper belt. The probe was launched from Cape Canaveral on January 19, 2006. It flew past Jupiter in February 2007 to take advantage of the planet's gravity and increase its speed. It made its closest approach to Pluto on July 14, 2015. Finally, the probe turned to one or more objects in the Kuiper belt. The principal objectives of the mission are to study the form and structure of Pluto and its satellite Charon, analyze the variability of the temperature on Pluto's surface, look for additional satellites around Pluto, and obtain high-resolution images. The power source for the spacecraft is a radioisotope thermoelectric generator.

**RADIOISOTOPE GENERATOR**
provides energy for propulsion of the spacecraft.

**LOW-GAIN ANTENNA**
Auxiliary to the high-gain antenna, which it can replace in case of breakdown.

**LAUNCH**
**January 19, 2006**
The New Horizons probe is launched from Cape Canaveral toward Jupiter, Pluto, and the Kuiper belt.

**JUPITER FLYBY**
**February 2007**
The probe flies by Jupiter to take advantage of the gravity of the planet on its journey toward Pluto.

**KUIPER FLYBY**
**2016-2020**
The probe flies by one or more Kuiper belt objects.

**INTERSECTING THE ORBIT OF MARS**
**April 7, 2006**
The probe traverses the Martian orbit.

**ARRIVAL AT PLUTO**
**July 14, 2015**
New Horizons flies by Pluto and its moon Charon. It sends to Earth data about the surface, the atmosphere, and the climate.

## THE SPACECRAFT

The central structure of New Horizons is an aluminum cylinder that weighs 465 kg (1,025 pounds), of which 30 kg (66 pounds) are accounted for by scientific instruments. All its systems and devices have backups. The spacecraft carries a sophisticated guidance-and-control system for orientation. It has cameras to follow the stars and help find the right direction. These cameras have a star map with 3,000 stars stored in their memory. Ten times each second, one of the cameras takes a wide-angle image of space and compares it with the stored map.

**TECHNICAL SPECIFICATIONS:**

| | |
|---|---|
| Date of launch | January 19, 2006 |
| Cost | US $650 million |
| Closest approach | 465 kg (1,025 pounds) |
| Organization | NASA |

0.7 m (2.5 ft)

2.1 m (7 ft)

**ANTENNA**
High-gain, 2.2 m (7 ft) in diameter, its purpose is to communicate with the Earth.

**RADIOMETER**
measures the atmospheric composition and temperature.

### 2015
**THE SPACECRAFT NEW HORIZONS ARRIVED AT PLUTO ON JULY 14.**

**TELESCOPIC CAMERA**
will map Pluto and gather high-quality geologic data.

**THRUSTERS**
The spacecraft carries six thrusters to increase its speed during flight.

## THE VENUS EXPRESS MISSION

Venus is a little smaller than the Earth and has a dense atmosphere. Because it is located at slightly more than 108 million km (67 million mi) from the Sun, it receives almost twice the solar energy as the surface of the Earth. The Venus Express is the first mission of the European Space Agency to Venus. The scientific aims include studying in detail the atmosphere, the plasma medium, the surface of the planet, and surface-atmosphere interactions. It was launched from the Baikonur Cosmodrome on November 9, 2005. The spacecraft entered into orbit on April 11, 2006, and the mission lasted until December 2014.

**TECHNICAL SPECIFICATIONS:**

| | |
|---|---|
| Launch | November 9, 2005 |
| Cost | US $260 million |
| Weight | 1,240 kg (2,700 pounds) |
| Organization | ESA |

6 feet (1.8 m)

5 feet (1.5 m)

**SPECTROMETER 1**
Measures the atmospheric temperature.

**SPECTROMETER 2**
Operates on ultraviolet rays.

**SOLAR PANELS**
Capture the energy from the Sun that powers the mission.

**CAMERA**
Captures images in the ultraviolet.

**MAGNETOMETER**
Measures magnetic fields and their direction.

**HIGH-GAIN ANTENNA**
Transmits data to Earth.

**LAUNCH**
November 9, 2005.

**ARRIVAL AT VENUS**
April 11, 2006.

**STAY ON VENUS**
Until December 2014.

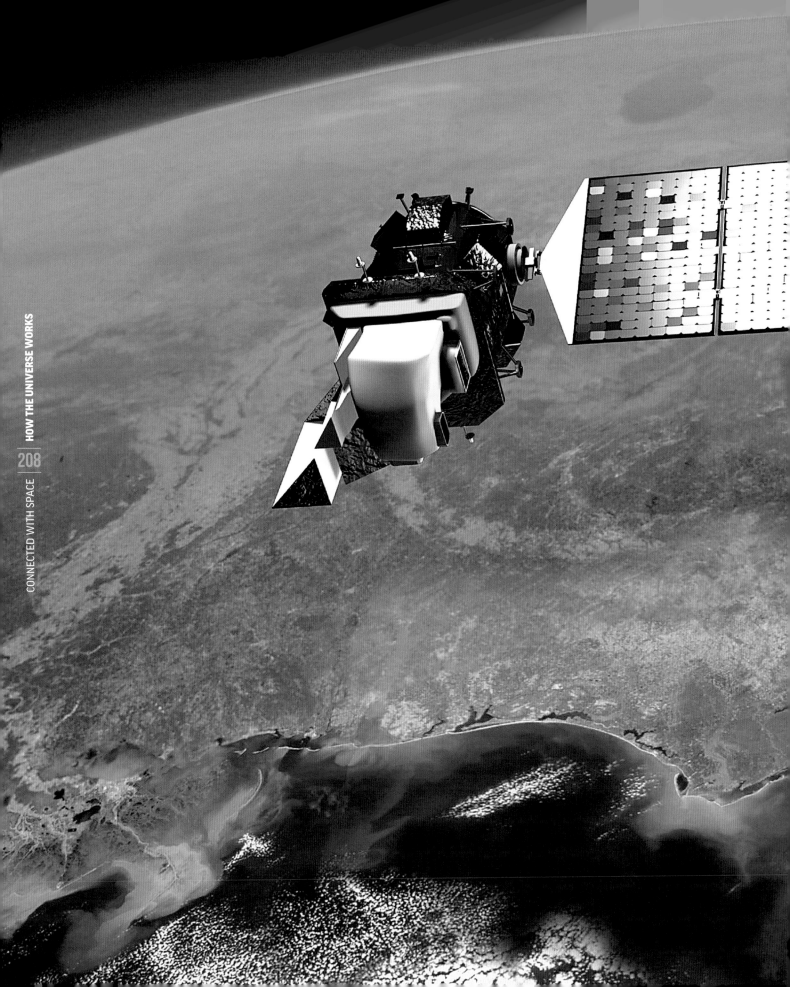

# CHAPTER 8

# CONNECTED
# WITH SPACE

In addition to providing answers about the Universe and the origin of life, space exploration has also represented great benefits for life on Earth. Many of the technological advances have been applied to fields such as medicine, safety or transportation. Communication, meteorological, observation and navigation satellites are another clear example. In recent years, space has also become the setting for incredible adventures and challenges, such as the Red Bull Stratos mission.

# SPACE TECHNOLOGY AT HOME

**Space has served as a research and development laboratory for new technologies and methods, the application of which is reflected in our daily lives.** Various devices, types of food, clothing, materials and utensils have been tested in space under extreme conditions and have improved the quality of our lives.

## INTELLIGENT CLOTHING

Clothing featuring computers and other technological elements are already a reality. Electronics can transform clothing into an intelligent biometric suit that responds to the environment in which the wearer is located and can measure his/her vital signs. Thanks to new fabrics, scientists are now discussing the development of garments to prevent illnesses.

### MAMAGOOSE

Mamagoose pyjamas can monitor babies while they sleep. They are equipped with sensors that monitor the baby's heartbeat and breathing. The pyjamas detect and provide a warning of possible sudden infant death syndrome symptoms. The vital signs of astronauts are controlled using a similar system.

**FIVE SENSORS**
Three on the chest
Two on the stomach

### POLYCARBONATE

Compact polycarbonate sheets are used in the construction industry. They have a high impact strength, and have replaced glass in some applications, including goggles.

## DOMESTIC USES

The popularization of space travel has resulted in the introduction of new technologies in our homes, such as microwave ovens and dried foods. Both have relatively recently taken a place in the daily lives of families at home.

### VELCRO
A quick-open and close system, created by George de Mestral in 1941.

### FOODS
Explorers have dry food stored in cool places. The menu includes dried fruits, smoked turkey, flour tortilla, soy milk cheese and nuts.

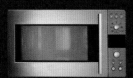

### MICROWAVE OVENS
Made popular in the USA in the 1970s, food can be cooked or quickly reheated thanks to the use of electromagnetic waves.

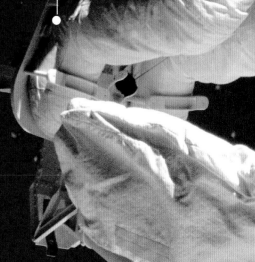

and are beneficial for allergy and asthma sufferers. They are mobile and can be moved from one room to another.

The purifier takes in air contaminated with allergens.

A filter processes the contaminated air.

The purifier returns pure air to the room.

Contaminated air

Pure air

## PROTECTED CRAFT

To withstand the effects of extreme temperatures and impacts against meteorites, spacecraft must be protected by various layers. The outer shell is made of aluminium, which covers a screen for protection against high temperatures. The inside is fitted with a screen to provide protection against low temperatures. A layer of adhesive silicone is responsible for joining them together.

**Screen for high temperatures**
Offers protection against the adverse effects of the Sun.

**Screen for low temperatures**
Offers protection from extreme cold temperatures.

**Silicone adhesive**

**Aluminium**
To protect the spacecraft against impacts with meteorites.

### KEVLAR
Synthesized polyamide, used in clothing for which resistance is essential, such as equipment for outdoor sports, bulletproof jackets and covers.

### SILICONE
Many polymers are made from silicone: it is used in lubricants, waterproof adhesives, kitchen molds and medical devices.

### TEFLON
The common name for polytetrafluoroethylene. Its special characteristic is that it is almost inert and does not react with other chemical substances, except under very special circumstances. It is also renowned as being waterproof and non-stick. It is used in the lining of rockets and planes, and at home, in pots and pans.

western union

MCDONNELL DOUGLAS

WESTAR VI

HUGHES
HUGHES AIRCRAFT COMPANY

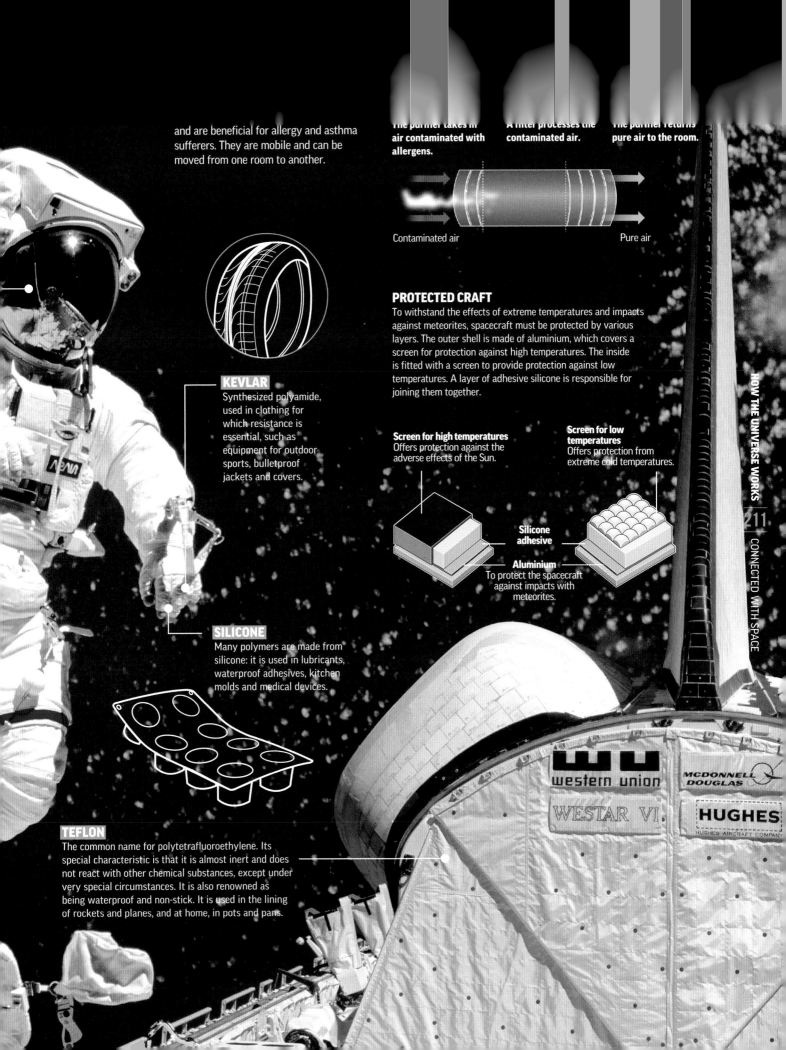

# MICROGRAVITY AND SCIENCE

**The microgravity that exists in space is an inconvenience to the health of astronauts because it produces, among other things, a decrease in heart rate, muscle weakening and calcium loss in bones.** The upside is that these special conditions of low gravity are conducive to scientific experiments. That is why the different laboratories of the International Space Station (ISS) have been working for years on advances that have a direct application to life on Earth.

## ADVANTAGES OF MICROGRAVITY

Under microgravity conditions, many materials crystallize differently than they do on Earth under the force of gravity. Then these elements react and behave in a way hitherto unknown on our planet, which offers new possibilities in fields as diverse as electronics, medicine or transport. Also in biology, because the cells and tissues of organisms do not grow the same in microgravity.

### ISS LABORATORIES

① **Kibo**
The Japanese experimental module is the largest of the ISS. It has a pressurized zone to perform microgravity experiments.

② **Columbus**
The ESA pressurized laboratory, cylindrical, is mainly used to study the science of materials and the physics of fluids.

③ **Destiny**
Since 2001, the NASA experimental module has been dedicated to the study of materials, physics, biotechnology, engineering and medicine.

**LEGS**
During weightlessness, an astronaut's legs get thinner from lack of exercise, and the muscles atrophy.

# PARABOLIC FLIGHT

On Earth it is also possible to achieve microgravity conditions. Parabolic flights allow astronauts to simulate space flights and conduct scientific experiments. To achieve microgravity, a C-135 aircraft ascends at an angle of 47° until the pilot shuts off the engines and the plane begins its free fall by following a parabolic trajectory. During this phase, everything in the airplane floats, both equipment and people, because they are in a weightless condition.

**LIQUIDS**
disperse in the air during conditions of microgravity.

0 g

**8,500 M (5.3 MI)**
The engines are stopped.

The engines are started again.

**7,600 M (4.7 MI)**
The engine velocity decreases.

**6,000 M (3.7 MI)**
Acceleration of the engines.

## 31
**FLIGHTS PER SESSION.**

1.8 g    1.8 g/1.5 g    **MICROGRAVITY**    1.8 g

**WITHIN THE AIRPLANE**
During parabolic flights, nine to 15 scientific experiments can be performed.

## 15
**EXPERIMENTS PERFORMED.**

**EXPERIMENT 8**
Studying smell and taste.

**EXPERIMENT 7**
Testing a new shower system for the astronauts.

**EXPERIMENT 15**
The behavior of ferro-fluids.

**FREE-FLIGHT ZONE**

## WEIGHTLESSNESS

Astronauts seem to float on their ships or on the ISS as a result of the free fall effect created by orbiting the Earth. It is what is known as microgravity.

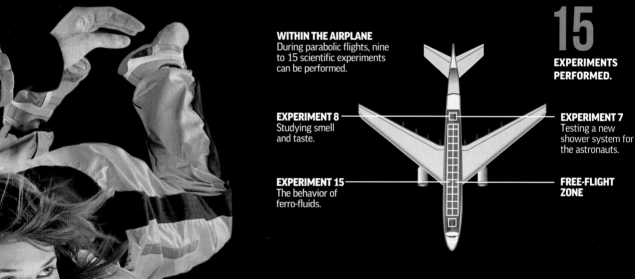

## WATER: ANOTHER METHOD FOR SIMULATING WEIGHT REDUCTION

Other methods of training for astronauts make use of a gigantic swimming pool in which an environment can be created to simulate working in microgravity. At the Johnson Space Center, a fully underwater simulator was installed that allows astronauts to test work as if they were outside the ISS. For this they use special suits with neutral buoyancy.

# GLOBAL INTERCONNECTION

Communications using satellites have made it possible to connect places that are very far from one another and to bring information to very remote regions. The satellites are primarily in geosynchronous orbits—that is, the satellite orbits in the same time it takes the Earth to rotate. This motion allows for more effective transmission systems, because the satellite is stationary with respect to the Earth's surface. There is a virtual fleet of geosynchronous satellites dedicated to various goals: meteorology, research, navigation, military uses, and, obviously, telecommunications.

## CONNECTIONS

Communication can be established between any two points on Earth. The signals sent and received between terrestrial and satellite antennas are in the radio-wave spectrum, and they range from telephone conversations and television to computer data. A call from Europe to the United States, for example, involves sending a signal to a terrestrial station, which retransmits the signal to a satellite. The satellite then retransmits the signal so that it can be received by an antenna in the United States for transmission to its final destination.

**DOWNWARD LINK**
The satellite retransmits signals to other points: a downward connection is made.

**UPWARD LINK**
The satellite captures signals that come from the Earth. An upward connection is made.

## TERRESTRIAL STATIONS

These stations are buildings that house the antennas and all the necessary equipment on land for sending and receiving satellite signals. The buildings can be large structures, and the antennas can act as receivers and transmitters of thousands of streams of information. In other cases, they are small buildings equipped for communications but designed to operate on board ships or airplanes.

**TRANSMITTING ANTENNA**
The terrestrial antenna receives information from the satellite and retransmits it. This is the key for every kind of telecommunication.

**TELEVISION BROADCAST CONNECTIONS**
Makes it possible to transmit the news or other events via satellites that capture the signals and distribute them to different geographic locations.

**NATIONAL TRANSMISSION GRID**
Fixed structure on Earth that communicates with the antenna and receives information

**PUBLIC NETWORK**
For telephone communication between two points

**PRIVATE NETWORK**
Groups of private corporations, such as TV networks

**PRIVATE**
Private clients who pay for satellite access

**MOBILE UNIT**
Used for covering news or events that occur in different locations

## FIXED SENDING AND RECEIVING ANTENNA

These antenna can target specific places on the Earth.

## TRANSPONDER

This is the heart of the satellite. It corrects for atmosphere-produced distortions of the radio signals.

## SOLAR PANELS

Takes the light from the Sun and transform it into electrical energy.

## REFLECTOR

Captures signals and retransmits them directly.

REFLECTOR

## MOVEMENT ON THREE AXES

To correct its position, the satellite turns in three directions: an axis perpendicular to its orbit and the horizontal and vertical axes.

PITCH

Direction to the Earth

ROLL

Velocity Vector

YAW

Orbit

## CONSTELLATION OF IRIDIUM SATELLITES

Iridium is a satellite-based, mobile telephone system in low Earth orbit. It consists of 66 satellites that follow a polar orbit.

## TELEPHONE COMMUNICATIONS

Makes communications possible between an airplane and land by means of satellites.

## TELEPHONY CONNECTION

A terrestrial antenna receives signals and transmits them to a center that resends to them in the corresponding format.

CENTER/ OPERATOR

Area of Maximum Power

Low-Power Boundaries

## SATELLITE FOOTPRINT

Transmitted radio waves cover a defined area when they arrive at Earth. The area is known as a footprint.

**TV**
The signal arrives from the center via the antenna

**LAND LINE**
The voice signal goes from the center to the desired location

**MOBILE TELEPHONES**
Can receive voice and images depending on the signal sent

# SATELLITE ORBITS

**The space available for placing communications satellites is not unlimited.** On the contrary, it is a finite space that could become saturated with too many satellites. Errors of 1 or 2 degrees in terms of location may cause interference between neighboring satellites. Therefore, their positions are regulated by the International Telecommunication Union (ITU). Geostationary satellites (GEO) maintain a fixed position relative to Earth. In contrast, those in low (LEO) and medium (MEO) Earth orbit require monitoring from ground stations.

## DIFFERENT TYPES

The satellites transmit information of a given quality according to their position in relation to Earth. GEO orbits can cover the entire Earth with only four satellites, while lower orbits like the LEO need constellations of satellites to have full coverage. In other cases, as in the MEO, satellites follow elliptical orbits.

| ORBITS | LEO | MEO | GEO |
|---|---|---|---|
| Distance from Earth | 200–3,000 km (125–1,865 miles) | 3,000–36,000 km (1,865–22,400 miles) | 36,000 km (22,400 miles) |
| Cost of satellites | Low | Medium | High |
| Type of network | Complex | Medium | Simple |
| Life of satellite | 3–7 years | 10–15 years | Simple |
| Coverage | Short | Medium | Continuous |

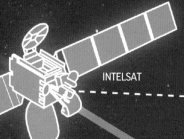

INTELSAT

36,000 km (22,400 mi) altitude

Polar Orbit

## GEO ORBIT

The geostationary orbit is circular and the most often used. The period of the orbit is 23 h 56 min, the same as that of the Earth. Its most frequent use is for television.

## LEO ORBIT

Low-altitude, between 200 and 3,000 km (125 and 1,865 mi). Was first used in cellular telephony after the saturation of GEO orbits. The orbits are circular and consume less power, although they do require terrestrial centers to track the satellites.

## ELLIPTICAL ORBIT

**Apogee**
Farthest point from Earth.

**Perigee**
Closest point relative to Earth.

## CIRCULAR ORBIT

**Same distance**

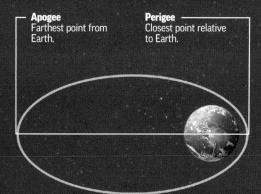

## 36,000 KM
### (22,400 MI)

THIS IS THE REQUIRED DISTANCE OF A SATELLITE'S ORBIT SO THAT IT REMAINS FIXED WITH RESPECT TO EARTH.

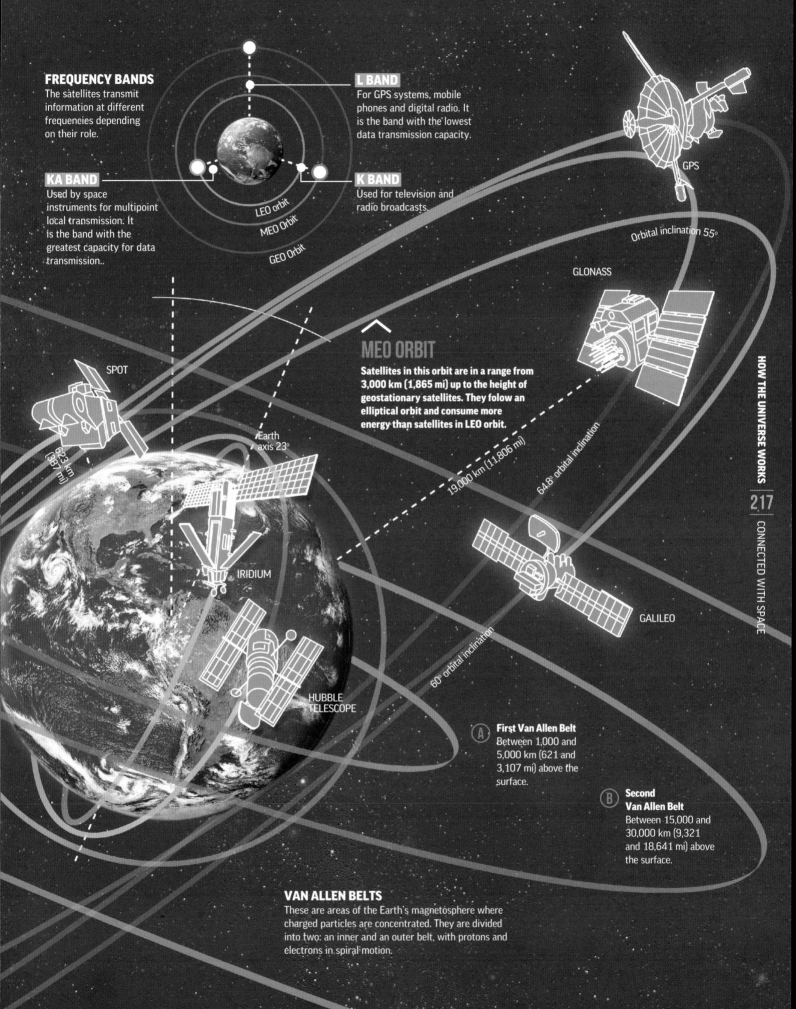

## FREQUENCY BANDS
The satellites transmit information at different frequencies depending on their role.

### KA BAND
Used by space instruments for multipoint local transmission. It is the band with the greatest capacity for data transmission..

### L BAND
For GPS systems, mobile phones and digital radio. It is the band with the lowest data transmission capacity.

### K BAND
Used for television and radio broadcasts.

LEO orbit

MEO Orbit

GEO Orbit

GPS

Orbital inclination 55°

GLONASS

SPOT

623 km (387 mi)

Earth axis 23°

## MEO ORBIT
Satellites in this orbit are in a range from 3,000 km (1,865 mi) up to the height of geostationary satellites. They follow an elliptical orbit and consume more energy than satellites in LEO orbit.

19,000 km (11,806 mi)

64.8° orbital inclination

IRIDIUM

GALILEO

60° orbital inclination

HUBBLE TELESCOPE

**(A) First Van Allen Belt**
Between 1,000 and 5,000 km (621 and 3,107 mi) above the surface.

**(B) Second Van Allen Belt**
Between 15,000 and 30,000 km (9,321 and 18,641 mi) above the surface.

## VAN ALLEN BELTS
These are areas of the Earth's magnetosphere where charged particles are concentrated. They are divided into two: an inner and an outer belt, with protons and electrons in spiral motion.

# ENVIRONMENTAL SATELLITES

Under the guidance of the French Space agency CNES, Spot 1 was put into orbit in 1986; it was the first satellite of what is today a satellite constellation that can take very high-resolution photographs of Earth. The latest version, Spot 7, was launched in 2014. Today, it is considered a commercial satellite par excellence, used by companies in the oil and agricultural industries. The USA, in turn, launched Landsat in 1972, the latest version of which was launched in 2017.

## SPOT 7 CAPABILITIES

The development of the Spot satellite constellation made it possible to commercially offer photographic monitoring of events linked to the environment. The latest version, the SPOT 7, has two NAOMI (New AstroSat Optical Modular Instrument) cameras capable of obtaining a resolution of 2.2 m (7.2 ft) in panchromatic mode, a resolution that rises to 1.5 m (4.9 ft) after being processed. Each image occupies a width of 60 km (37 mi). The constellation formed by the SPOT 6 and SPOT 7 satellites is able to photograph about six million square kilometers of the Earth's surface per day and works in coordination with the two French Pleiades observing satellites.

### SPOT SATELLITES
They work together, thanks to which it is possible to obtain an image from any point of the globe daily.

### TECHNICAL SPECIFICATIONS

| | |
|---|---|
| Launch date | June 30, 2014 |
| Orbital latitude | 407 mi (655 km) |
| Orbital period | 98.79 minutes |
| Maximum resolution | 1.5 m (4.9 ft) |
| Organization | Airbus Defence & Space |

1.55 m (5 ft)

1.75 m (5.7 ft)

## U.S. SATELLITES

Landstat 8 has two terrestrial observing instruments: the Operational Land Imager (OLI) and the Thermal Infrared Sensor (TIRS). OLI and TIRS will collect the data together to provide coincident images of the Earth's surface. It will be located in a heliosynchronous polar orbit 705 km (438 mi) high with a slope of 98.2°. From this orbit you can observe the entire surface of the Earth every 16 days.

### LANDSAT 8
Built by Orbital Sciences Corporation in Gilbert, Arizona, has a shelf life of 5 years, but carries enough fuel for 10 years of operations.

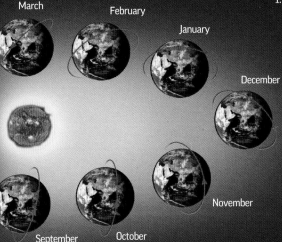

March
February
January
December
November
October
September

## SUN-SYNCHRONOUS ORBIT

In order to compare observations of a given point captured on different dates, the images must be taken under similar light conditions. To this end, a Sun-synchronous orbit is used, which means it is possible to view the entire Earth's surface for a period of 26 days.

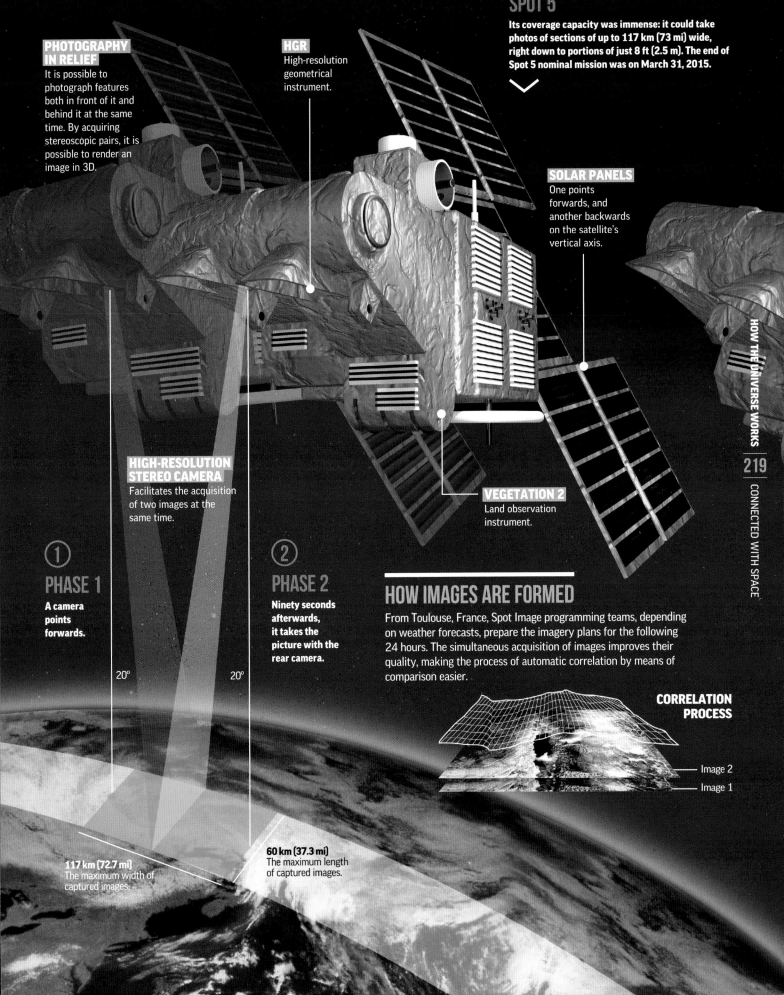

**PHOTOGRAPHY IN RELIEF**
It is possible to photograph features both in front of it and behind it at the same time. By acquiring stereoscopic pairs, it is possible to render an image in 3D.

**HGR**
High-resolution geometrical instrument.

SPOT 5
Its coverage capacity was immense: it could take photos of sections of up to 117 km (73 mi) wide, right down to portions of just 8 ft (2.5 m). The end of Spot 5 nominal mission was on March 31, 2015.

**SOLAR PANELS**
One points forwards, and another backwards on the satellite's vertical axis.

**HIGH-RESOLUTION STEREO CAMERA**
Facilitates the acquisition of two images at the same time.

**VEGETATION 2**
Land observation instrument.

① **PHASE 1**
A camera points forwards.

② **PHASE 2**
Ninety seconds afterwards, it takes the picture with the rear camera.

20°     20°

## HOW IMAGES ARE FORMED

From Toulouse, France, Spot Image programming teams, depending on weather forecasts, prepare the imagery plans for the following 24 hours. The simultaneous acquisition of images improves their quality, making the process of automatic correlation by means of comparison easier.

**CORRELATION PROCESS**

Image 2
Image 1

**60 km (37.3 mi)**
The maximum length of captured images.

**117 km (72.7 mi)**
The maximum width of captured images.

# EARTH IMAGES IN HIGH RESOLUTION

**The images taken by satellites like Spot make it possible to view the relief of any region on the planet at different scales,** from photographs that capture 2.5 m (8 ft) of terrain, to sections up to 60 km (37.3 mi) wide. The powerful definition of Spot makes it possible to home in on very specific targets, from vegetation to port areas, from seas and geographical boundaries to fire areas.

## IMAGE RESOLUTION

The Spot satellites obtain a maximum definition of 1.5 m (4.9 ft) on Earth, a high resolution that makes it possible to cover very specific aspects of the target areas and mapping anywhere in the world on a 1:25,000 scale. The images provided by Spot are used to control harvests, prevent natural disasters and observe demographic growth.

**ISRAEL**
**Latitude** 32.98°
**Longitude** 35.57°

| SATELLITE | PIXEL SIZE | IMAGE |
|---|---|---|
| Spot 1 to 3 | 10 m (33 ft) | Color and B&W |
| | 20 m (66 ft) | Color |
| Spot 4 | 5 m (16 ft) | Color and B&W |
| | 10 m (33 ft) | Color |
| Spot 5 | 2.5 m (8 ft) | Color or B&W |
| | 5 m (16 ft) | Color or B&W |
| | 10 m (33 ft) | Color or B&W |
| Spot 6-7 | 1.5 m (4.9 ft) | Color or B&W |

## COMBINED

If used together, Spot 6 and 7 satellites can revisit any point on the planet daily and cover up to 6 million km² (2.3 million square mi) each day.

## 3D IMAGERY

**THE SCANNING METHOD USED BY SPOT MAKES IT POSSIBLE TO BUILD IMAGES IN THREE DIMENSIONS, FOR ALL TYPES OF TERRAIN.**

## 10 YEARS

**IS THE USEFUL LIFE OF SPOT SATELLITES. DEVICES 6 AND 7 WILL RUN UNTIL 2024.**

Haifa

ISRAEL

Gaza

SEA OF GALILEE

● Nazareth

**3,600 KM²**
**(1,390 SQ MI)**

**THE MAXIMUM SURFACE AREA
THAT CAN BE COVERED IN AN IMAGE
TAKEN BY SPOT 6-7. IMAGES CAN
BE TAKEN AT BOTH A LOCAL SCALE
(FOR WHICH A FINER RESOLUTION IS
USED) AND AT A REGIONAL SCALE.**

**SYRIA**

## WEST BANK

This is one of the most
densely populated areas
on Earth. Its characteristic
desert landscape can be
seen in this real-life color
photograph taken by
Spot 5.

## LANDSAT 7 IMAGES

The US satellite took this photograph of the Dead
Sea in February 1975. The image combines optical
techniques and infrared techniques (at this wave
range, water appears black). The Dead Sea is at the
center, flanked by Israel and Jordan.

**Desert**
Seen here as brown.

**Vegetation**
Seen here as green.

## JUDEAN DESERT

The photograph of the
Judean wilderness shows
the different elevations at
an impressive resolution.
The region of Sodom,
387 m (1,270 ft) beneath
sea level, is the lowest
place on the planet.

River Jordan

Jerusalem ●

## THE DEAD SEA

The lowest body of water on Earth, 400 m
(1,312 ft) below sea level. Water quickly
evaporates in this desert climate. It leaves
behind remains of dissolved minerals.

DEAD SEA

# SPACE JUNK

**Since the launch of the first satellite in 1957, the space around Earth has become littered by a huge amount of debris.** Spent satellite batteries, parts of rockets and spacecraft orbit around Earth. The danger of these objects' presence is the possibility of a collision: they travel at speeds ranging from 30,000 to 70,000 km/h (18,640 to 43,500 mph).

## COSMIC RUBBISH

Any useless, artificial object that orbits Earth is considered space debris. Rockets used just once continue orbiting the planet, just like bits of spacecraft or devices intentionally destroyed so that they do not move into incorrect orbits.

### THE SIZE OF SPACE DEBRIS

More than 11,000 catalogued objects have been accumulated, in addition to millions of tiny particles.

**Measure less than 1 cm (0.4 in)**
Very small particles cause limited surface damage.

+30,000,000

**Measure between 1 cm and 10 cm (0.4 in and 4 in)**
Particles that can create holes in satellites.

+100,000

**Measure more than 10 cm (4 in)**
Capable of causing irreparable damage. They have been catalogued and are monitored from Earth.

+11,000

**17,385**

**THE NUMBER OF OBJECTS CURRENTLY IN ORBIT (2016).**

## OBJECTS IN SPACE BY COUNTRY

Since 1957, 25,000 objects have been launched into low orbit. Most come from Russia and the United States.

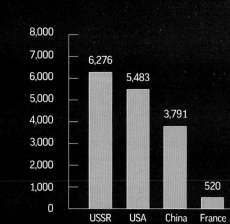

| Country | Objects |
|---|---|
| USSR | 6,276 |
| USA | 5,483 |
| China | 3,791 |
| France | 520 |
| Japan | 215 |
| India | 176 |
| ESA | 109 |
| Others | 815 |

## WHAT CAN WE DO?

One solution could be to return all debris to Earth, rather than leaving it orbiting around the planet. However, what has been achieved is to work on satellite debris to remove it from Earth's orbit.

**SAIL**

Just like on a boat, the sail is released when the satellite stops working and the solar winds divert it.

**SPACE PROBE**

It impacts against the satellite, which is diverted from its orbit and driven in a pre-established direction.

**CABLE**

A cable drags the satellite to lower orbits. It disintegrates when entering the atmosphere.

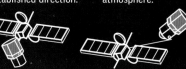

### SOURCE AND LOCATION

95 per cent of objects in space around Earth are considered 'space waste'. NASA is studying a way of using rockets that do not reach orbit and that will return to Earth, preventing the creation of more waste.

**21 per cent** inactive satellites

**5 per cent** active satellites

**31 per cent** rockets and rocket boosters

**43 per cent** fragments of satellites

## 2,000

**TONNES OF RUBBISH AT LESS THAN 2,000 KM (1,243 MILES).**

**POLAR ORBIT**
**400 KM (249 MI)**
The ISS and the Hubble telescope orbit here.

**LOW ORBIT**
**700-2,000 KM (435-1,243 MI)**
Telecommunication and environmental satellites.

**ORBIT GEOSTATIONARY**
**35,800 KM (22,245 MI)**
Spy satellites, which generate a significant amount of waste.

**HIGH ORBIT**
**100,000 KM (62,137 MI)**
Astronomic satellites operate at the highest orbit.

■ Waste
▨ Functional
▦ Nuclear spills

# GLOBAL SATELLITE NAVIGATION

GALILEO

**The Global Positioning System (GPS) was developed by the US Department of Defense and makes it possible to establish the position of a person, vehicle or spacecraft anywhere in the world.** To achieve this, it uses a constellation of two dozen NAVSTAR satellites. It was fully deployed in 1995 and, although it was designed for military use, it is now applied in various fields and forms part of our daily lives. The European Union is developing the Galileo System, similar to GPS, but comprising 30 satellites.

① **PHASE 1**

The first satellite sends its coordinates. The navigational aid captures the signal, indicating its distance from the satellite within the scanned sphere.

② **PHASE 2**

If a second satellite is added, an area is established within the intersection of both spheres in which the navigational aid is found.

SATELLITE A

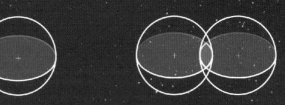

SATELLITE A

SATELLITE B

COVERAGE AREA

## FUNCTIONALITY

Using the electromagnetic waves sent by the satellite, receivers can convert the signals received into estimate positions, speeds and times. To calculate an exact position, four satellites are required. The first three form a tripartite area of intersection, while the fourth works to correct the position. When the area scanned by the fourth satellite does not coincide with the previously established intersection, the position is corrected.

# GALILEO SYSTEM

The European Galileo project (whose first experimental satellite was put into orbit in 2005) is a satellite-based navigational system that will use a set of 30 satellites. These will orbit Earth at 23,000 km (14,300 mi), on three different planes to offer complete coverage. It is still not operational, but it is expected to be fully functional by 2020.

3 m (10 ft)

## ELECTROMAGNETIC WAVES

Using the electromagnetic waves sent by satellites, the receiver calculates the distance and position of the point sought. The waves travel at 300,000 km/sec (186,400 mi/sec).

SATELLITE A

SATELLITE C

SATELLITE B

## GALILEO'S ORBIT

The satellites' orbits ensure there is sufficient coverage to precisely calculate positions on Earth.

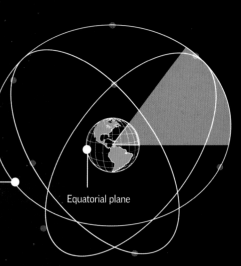

### ORBIT
Around 55° on the equatorial plane.

Equatorial plane

③
## PHASE 3

**Combining the three satellites, a common point can be established that indicates the exact position of the navigational aid.**

## GALILEO SATELLITE

| First launch | 2005 |
|---|---|
| Orbital height | 23,000 km (14,300 mi) |
| Orbital period | 14 hours |
| Organization | European Union |

SATELLITE A

SATELLITE B

④
## PHASE 4

**A fourth satellite is required to correct any possible positional error.**

SATELLITE C

SATELLITE D

## THE RECEIVER

It is equipped with all the controls required to precisely establish the location of a given point. It provides the user with all the coordinates needed.

### INDICATOR
of latitude, longitude and height.

### CONTROL
To utilize the device's map.

Keep Right at Main Street

1/2 mi

1:30
15.5 mi
eta 13.17

11:47

500 ft

GPS

POINT OF RECEPTION

# SPACE VACATIONS

In 2001, US multimillionaire Dennis Tito, the first space 'tourist,' was successfully sent to the International Space Station; he paid US $20 million for an eight-day stay. Australian Mark Shuttleworth followed in his footsteps in 2002. Later, SpaceShipOne, designed by Burt Rutan, and its successor, SpaceShipTwo (Virgin Galactic), were created to send thousands of tourists to traveling into space in the coming years.

## THE JOURNEY

SpaceShipOne is propelled by the plane White Knight. The suborbital flight lasts approximately two hours, at a maximum speed of 3,580 km/h (2,225 mph) and a maximum height of 100 km (62 mi). The stay in outer space lasts six minutes, from where the traveller can enjoy the globe's profile and experience the effects of low gravity.

35 m (115 ft)

82 m (269 ft)

**MAXIMUM HEIGHT**
The spaceplane reaches a height of 100 km (62 mi), before falling back into the atmosphere. The crew experiences low gravity for six minutes.

**5,000 KG**
**(11,000 LB)**
WEIGHT ON THE GROUND.

**3,670 KG**
**(8,090 LB)**
THE WEIGHT OF SPACESHIPONE.

### WHITE KNIGHT PLANE

| | |
|---|---|
| Launch date | June 2004 |
| Maximum height | 15.24 km (9.5 mi) |
| First pilot | Mike Melvill |
| Company | Private |

Height (km)

100

90

80s

70

60

50

40

30

**THE ENGINE**
is ignited for 80 seconds and reaches a speed of 3,580 km/h (2,225 mph).

**TAKEOFF**
After one hour in flight, at a height of 15.24 km (9.5 mi), the launch plane, White Knight, releases SpaceShipOne.

**LANDING**
The craft touches ground again.

**RE-ENTRY**
The pilot configures the descent.

**GLIDING**
The space plane descends towards the Earth's surface.

**US $250,000**
COST OF THE SUBORBITAL FLIGHT

**4 DAYS**
TRAINING

**2 HOURS**
FLIGHT DURATION

## CREW

Seated at the rear of the plane. They wear pressurized suits and have undergone a rigorous training routine.

### SPACESHIPONE

| | |
|---|---|
| Launch date | June 2004 |
| Orbital height | 100 km (62 mi) |
| First pilot | Mike Melvill |
| Company | Private |

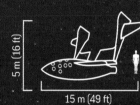

5 m (16 ft)

15 m (49 ft)

## THE CABIN

Equipped with cutting-edge technology, allowing the pilot to maneuver the plane safely. It is equipped with 16 circular glass panels that facilitate a panoramic view of space.

**Propellant**
Solid hybrid.

**Boosters**
Allow the plane to move up or down during flight.

**Engine**
housing with liquid fuel.

**Rudders**
Managed electrically, they provide longitudinal stability.

**Flaps**
Used to control the plane's height.

**Movement**
of the nose
From one side to the other, around the center of gravity.

**Feathers**
The wings and tail turn up to ensure a safe re-entry.

**CIRCULAR PANELS**
The 16 glass panels provide great views.

**HEIGHT GUIDE**
Used during re-entry into the atmosphere.

**RUDDER PEDALS**
While the plane turns, the rudder pedals prevent it from spinning around.

**DISPLAY**
Shows the position of the plane compared to Earth, the route to its destination and the amount of compressed air on the wings.

**CENTRAL LEVER**
Used to move the plane up or down.

**ENGINE**
Engaged using buttons, it burns fuel in 80 seconds.

**REGULATOR**
To control deviations from its flight path.

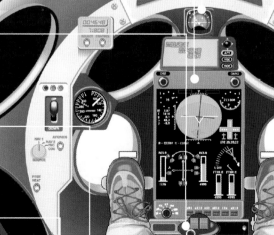

# STRATOSPHERIC ADVENTURE

**Taking off from the highest possible pitch is one of the most exciting experiences for any lover of adventure and risk.** The challenge is to ascend with a hot air balloon to the stratosphere, surpassing heights of 30,000 km, then descending vertiginously and landing successfully by opening a parachute as Earth is approached. This risky challenge involves overcoming extreme temperature changes, low atmospheric pressures and G forces that can cause uncontrolled body turns and loss of consciousness.

## JOSEPH KITTINGER, THE PIONEER

Although Austrian adventurer Felix Baumgartner famously jumped from more than 39,000 meters in an event that was televised worldwide, the true pioneer in stratosphere jumping was American colonel Joseph Kittinger. Fond of high-altitude parachute jumping, Kittinger promoted the Excelsior project and on August 16, 1960, completed a free fall from a helium balloon at a height of more than 31,000 meters above sea level.

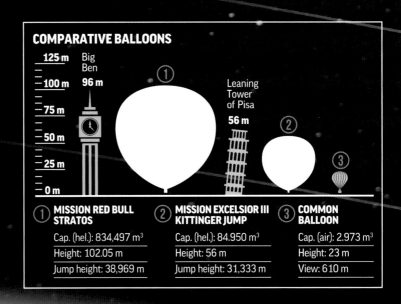

### COMPARATIVE BALLOONS

| | | |
|---|---|---|
| 125 m | Big Ben | |
| 100 m | 96 m | |
| 75 m | | Leaning Tower of Pisa |
| 50 m | | 56 m |
| 25 m | | |
| 0 m | | |

| ① MISSION RED BULL STRATOS | ② MISSION EXCELSIOR III KITTINGER JUMP | ③ COMMON BALLOON |
|---|---|---|
| Cap. (hel.): 834,497 m³ | Cap. (hel.): 84.950 m³ | Cap. (air): 2.973 m³ |
| Height: 102.05 m | Height: 56 m | Height: 23 m |
| Jump height: 38,969 m | Jump height: 31,333 m | View: 610 m |

## A NEW RECORD

It was less advertised, less televised and less expensive, but just as surprising. In October 2014, two years after Baumgartner's jump, another adventurer threw himself from the stratosphere and snatched the record from the Austrian. It was Alan Eustace, a recognized computer scientist and one of the vice presidents of computer giant Google. The 57-year-old executive boarded an enormous balloon inflated with helium over the New Mexico desert — a scenario similar to that chosen for the Red Bull mission — and launched into free fall from a height of 41,122 m (25.55 mi). He reached a speed of 1,320 km / h (820.2 mph) and, like his predecessor, broke the sound barrier.

### WITHOUT CAPSULE

Eustace took two hours to ascend the more than 40 km in height and 15 minutes to descend at full speed. Unlike Baumgartner, he did not travel within a sophisticated capsule; he did it inside a pressurized suit.

## SATELLITES
200-35,786 km
(125-22,235 mi)

## THE STRATOSPHERE

The atmosphere layer from which the large vacuum jumps have taken place extends up to 50 km (31 mi) in height. In this zone the temperature changes its tendency: it increases until reaching 0° C (32° C) in the stratopause (the boundary between the stratosphere and the mesosphere). It consists of a series of layers or layers and contains ozone, which acts as a filter to prevent the arrival of harmful radiation on Earth.

## ATMOSPHERE LAYERS

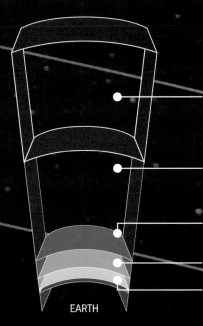

**EXOSPHERE**
Up to 10,000 km
(6,213 mi)

**THERMOSPHERE**
Up to 640 km
(397 mi)

**MESOSPHERE**
Up to 80 km
(49,7 mi)

**STRATOSPHERE**
Up to 50 km (31 mi)

**TROPOSPHERE**
Up to 9-18 km
(5.6-11.2 mi)

EARTH

## SPACE SHUTTLE
185-643 km
(115-400 mi)

# 10 TIMES

**THE HEIGHT OF BAUMGARTNER'S FREE FALL FROM THE STRATOSPHERE COMPARED TO A CONVENTIONAL PARACHUTE JUMP.**

**RED BULL JUMP**
39,000 m
(128,000 ft)

**JET PLANES**
25,900 m
(84,975 ft)

**COMMERCIAL AIRCRAFT**
12,500 m
(41,000 ft)

**PARACHUTE JUMP WITH INSTRUCTOR**
3,600 m
(11,800 ft)

**BALLOONS OR PROBES**
3,000 m
(9,850 ft)

On October 14, 2012 the Austrian adventurer Felix Baumgartner became an icon of extreme sport after jumping from a pressurized capsule carried into the stratosphere to a height of 39 km by a helium balloon. Besides being an historic event, Baumgartner's free fall allowed scientists to study the physiological consequences of a human body breaking the sound barrier.

## A SUPERSONIC JUMP

From the moment of the jump it took 34 seconds until Baumgartner reached the speed of sound. It took another 16 seconds to achieve top speed. In total, Baumgartner was at supersonic speed for half a minute. The complete free fall took four minutes and 20 seconds. After five years of preparation, the Austrian adventurer had completed a historic feat.

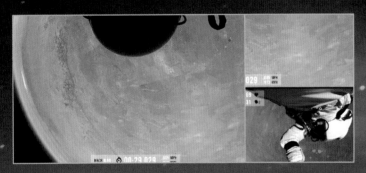

### REGISTERED FLIGHT

The multiple cameras attached to the suit allowed different images to be captured while the athlete's body was spinning uncontrollably and attempting to regain stability to open the parachute and return to solid ground.

### ADVENTURER

Passionate about extreme heights, Baumgartner began to practice skydiving at age 16. He entered a military sports club where he practiced free fall jumps. At 43 he achieved his stratospheric goal.

**39,000 m**
127,953 ft

**37,000 m**
121,391 ft

**34.000 m**
111,548 ft

**31.500 m**
103,346 ft

**27.000 m**
88,583 ft

**24.500 m**
80,380 ft

**21.000 m**
68,898 ft

**17.500 m**
57,415 ft

**14,000 m**
45,932 ft

**10,500 m**
34,449 ft

**7,000 m**
22,966 ft

**4,500 m**
14,764 ft

**1,500 m**
4,921 ft

**Sea level**

(2)

## JUMP ZONE

**The balloon reaches the doors of space in less than three hours. Baumgartner will jump from a height of approximately 39 km [24.23 mi].**

## -70⁰ C
### (-94⁰ C)

**AS IT RISES, TEMPERATURES ARE DROPPING, AND CAN REACH 70 DEGREES CELSIUS BELOW ZERO.**

(1)

## ELEVATION

**The helium inflated balloon which holds and elevates Felix Baumgartner's capsule is released.**

## THE CAPSULE

It consists of a spherical survival cell, similar to atmosphere reentry vehicles used in space missions. It is pressurized and equipped to guarantee Baumgartner's safe return to Earth in the event he is unable to make the jump. The survival cell is made of fiberglass and epoxy and is fireproof. Inside, the air consists of a mixture of and nitrogen that the traveler can regulate manually.

### ③ THE JUMP

Standing at the opening of the capsule, the extreme sportsman watches the edges of the Earth for a few seconds and then launches himself into space.

**STRATOSPHERE**

**TROPOSPHERE**

### ④ SOUND BARRIER

Baumgartner descends in free fall and breaks the sound barrier. The first part includes a period of uncontrolled spins.

## BAUMGARTNER RECORDS

# 38,969.4 M
## (127,852 FT)

**HIGHEST ALTITUDE OF A VACUUM JUMP.**

# 39,068.5 M
## (128,177 FT)

**HIGHEST ALTITUDE OF A MANNED BALLOON.**

# 1,357.6 KM/H
## (843 MPH)

**MAXIMUM VERTICAL SPEED (MACH 1.25).**

### ⑤ LANDING

At about 1,500 m (4,921 ft), the parachute is opened. Ten minutes later, Baumgartner lands safely.

### ⑥ LOCATION

The capsule and the balloon were recovered on land using a tracking device.

# INDEX

# INDEX

## A

**Aldrin, Buzz,** 149, 151
**Allen Telescope Array (ATA),** 180
**Allen, Paul,** 180
**ALMA (observatory),** 131
**Alouette I,** 145
**Ames Research Center,** 140
**Anomalocaris,** 105
**Apenninus (Moon mons),** 84
**Aphrodite Terra,** 53
**Apollo,** 141, 148, 149, 152-153, 158
  *8,* 149
  *9,* 141, 149
  *13,* 152, 153
  *14,* 148
  *17,* 152, 153
  *18,* 153
  *19,* 153
  *20,* 153
**Archaea,** 108
**Arecibo, observatory,** 180
**Ariane V (rocket),** 144, 145, 162, 163, 164, 165, 203
**Ariel (moon),** 61
**Aristarchus (crater),** 84
**Armstrong, Neil,** 149, 151, 169

**Arrhenius, Svante,** 110
**Asteroid,** 47, 68, 69, 145, 196
  *Hidalgo,* 69
  *Ida,* 69
  *Trojans,* 69
**Astrolabe,** 115, 116, 122
**Astronaut,** 142, 146, 147, 152, 160, 161, 168-169, 170, 171, 198, 210, 213
**Astronomy,** 114-115, 116-117, 120, 122, 202
**Atlas (rocket),** 148
**Atmosphere,** 47, 49, 50, 52, 53, 55, 56, 60, 62, 63, 65, 72, 78, 80, 81, 83, 91, 100, 131, 138, 167, 192, 200, 201, 202, 207, 231
**Atom,** 18, 19, 20, 23, 99
**Aurora Borealis,** 83

## B

**Baumgartner, Felix,** 228, 230, 231
**BEAM (module),** 173
**Becquerel, Paul,** 110
**Big Bang,** 18-19, 20, 21, 24, 25, 26, 30, 117, 129, 154, 155
**Big Crunch,** 24

**Big Ear (radio telescope),** 180, 181
**Biosphere,** 78
**Black Hole,** 25, 29, 30, 31, 35, 41, 42-43
**Blazar,** 31
**Brahe, Tycho,** 122
**Braun, Wernher von,** 138, 139

## C

**Callisto (moon),** 57, 201
**Calypso (moon),** 58
**Cambrian (period),** 91, 103, 104-105
**Canadian Space Agency (CSA),** 145
**Carbon,** 20, 36, 40, 41, 48, 92, 99, 167
  *Dioxide,* 52, 54, 55, 64, 91, 92, 96, 100
  *Monoxide,* 55, 65, 96
**Carboniferous (period),** 91, 107
**Cassini (probe),** 159, 200, 204, 205
**Cell,** 100, 101
**Cenozoic,** 92
**Ceres,** 69
**Cernan, Eugene,** 152
**Chandra (X-ray observatory),** 175, 176-177
**Charon (moon),** 64, 206
**Chromosphere,** 49
**Clementine (spacecraft),** 153
**Cluster,** 17, 27, 33, 175
  *Hercules,* 27
  *Omega Centauri,* 33
  *Pleiades,* 33
**Collins, Michael,** 151
**Columbia (orbital module),** 150, 151
**Columbus (laboratory),** 212
**Comet,** 70-71, 135
  *Halley,* 71
**Constellation,** 118-119, 132, 133, 134, 135
  *Aquarius,* 119
  *Aries,* 118
  *Cancer,* 118
  *Capricorn,* 119
  *Centaurus,* 119
  *Gemini,* 118

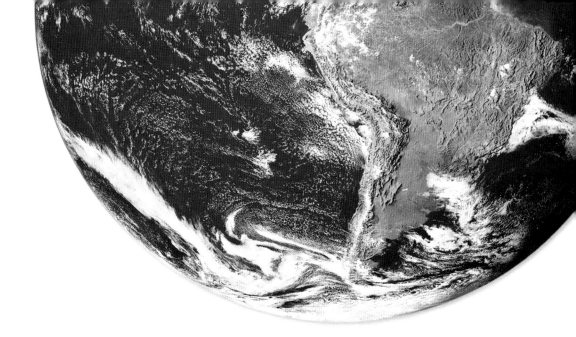

*Leo*, 118
*Libra*, 119
*Ophiuchus*, 119
*Orion*, 119, 134
*Pisces*, 118
*Sagittarius*, 119, 181
*Scorpio*, 119
*Taurus*, 118
*Ursa Major*, 119
*Virgo*, 119
**Continental drift,** 93
**Cooper, Gordon,** 148
**Copernicus (Moon crater),** 84
**Copernicus, Nicolaus,** 116, 117, 122
**Cretaceous (period),** 92
**Crisium (Moon sea),** 85
**Curiosity (rover),** 187, 194-195
**Cyclomedusa,** 103

# D

**Deep Impact,** 70
**Deep Space Network,** 176, 177
**Deimos (moon),** 54
**Destiny (laboratory),** 172, 212
**Devonian (period),** 91, 106
**Diamond,** 99
**Dickinsonia,** 103
**Drake, Frank,** 180, 181
**Dryden Research Center,** 140
**Dysnomia (moon),** 66

# E

**Ediacara,** 102
**Eagle (lunar module),** 150, 151
**Earth,** 16, 21, 26, 27, 32, 33, 36, 37, 38, 39, 46, 47, 51, 52, 53, 54, 56, 57, 58, 60, 62, 68, 69, 70, 74, 75, 78-79, 80-81, 82-83, 84, 86, 87, 88, 89, 90, 91, 92, 93, 94, 96, 97, 98, 100, 102, 110, 111, 116, 117, 118, 122, 124, 132, 133, 135, 146, 150, 152, 158, 167, 174, 175, 178, 179, 188, 189, 197, 200, 204, 207, 212, 213, 214, 218, 220, 222, 228
**Eclipse,** 88-89, 114, 126
**Ehman, Jerry R.,** 181
**Einstein, Albert,** 22, 23, 126, 127
**Electromagnetism,** 22, 23, 128, 225
**Electron,** 18, 19, 20, 23, 43, 71, 183
**Eocene (period),** 92
**Eris,** 67
**Eukaryote,** 100, 101, 108
**Europa (moon),** 57, 110, 200, 202-203
**Europa Clipper (probe),** 202-203
**European Space Agency (ESA),** 144, 153, 174, 175, 182, 197, 198, 202, 203, 204, 207
**Exomars,** 187, 198
**Exoplanet,** 72-73, 74
**Exosphere,** 81
**Explorer I (satellite),** 139, 146
**Extinction,** 91, 92
**Eustace, Alan,** 228

# F

**Filament,** 17, 20, 21, 41
**Fomalhaut b,** 73
**Friedmann, Alezander,** 127

# G

**Gagarin, Yuri,** 146, 147
**Galaxy,** 16, 17, 18, 19, 20, 21, 24, 25, 26-27, 28, 29, 30-31, 40, 41
*Andromeda*, 17
*Antennae*, 26
*Radio*, 31
*Sombrero*, 26
**Gale (crater),** 194
**Galileo (probe),** 159, 200-201, 217
**Galileo Galilei,** 28, 57, 117, 122, 123, 126, 138, 201, 202
**Gamow, George,** 25
**Ganymede,** 57, 201
**Gas Clouds,** 30, 31, 34
**Gemini (observatory),** 130
**Gemini,** 147, 148, 151, 153
*1*, 147, 148
*3*, 148
*4*, 153
*6*, 148
*7*, 148, 153
*8*, 151
*10*, 151
*12*, 153
*13*, 151
**Geocentrism,** 116, 122
**Glaciation,** 90, 93
**Gluon,** 18, 19, 23
**Goddard, Robert,** 138
**Goddard Space Flight Center,** 140
**Gold,** 41, 98, 183
**Gondwana,** 91, 92, 93
**GPS,** 217, 224-225
**Graphite,** 99
**Gravitation, Universal,** 22, 23, 117, 124
**Graviton,** 18
**Gravity,** 19, 20, 22, 24, 26, 30, 31, 42, 43, 47, 86, 126, 138, 169, 204, 212
**Guth, Alan,** 19
**Gyroscope,** 164

# H

**Haise, Fred,** 152
**Halite,** 98, 99

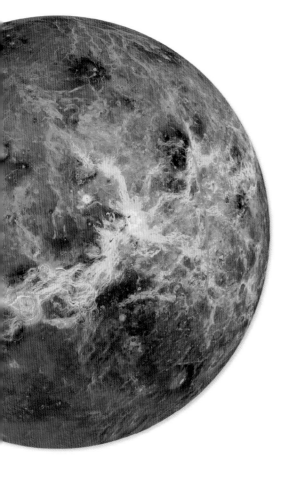

**Hallucigenia,** 105
**Ham (monkey),** 147, 148
**Hawking, Stephen,** 24
**HD 189733,** 73
**HD 20185 b,** 72
**HD 209458 b,** 72, 73
**Heliocentrism,** 117, 122, 123
**Heliopause,** 178
**Heliosphere,** 178, 182, 183
**Helium,** 19, 20, 23, 32, 36, 38, 48, 56, 57, 59, 60, 63, 228, 230
**Hertzsprung-Russell (diagram),** 32
**HiRISE (camera),** 191
**Holocene (period),** 93
**Homo Sapiens,** 21
**Hubble, Edwin,** 25, 26, 27, 117, 127
**Hubble (telescope),** 61, 64, 65, 72, 73, 75, 174-175, 217
**Humboldt (crater),** 85
**Huygens (probe),** 159, 204, 205
**Hydra (moon),** 64
**Hydrogen,** 19, 20, 23, 32, 34, 36, 37, 38, 39, 48, 56, 57, 59, 60, 63, 64, 71, 96, 100, 129, 166, 187, 188, 201
**Hydrosphere,** 81

## I

**Imbrium (Moon sea),** 85
**Inflation (Theory),** 18, 19

**InSight (Lander),** 187
**Inspiration Mars,** 199
**International Astronomical Union (IAU),** 64
**International Space Station (ISS),** 142, 145, 158, 166, 172-173, 196, 212, 213, 226
**Io,** 57, 200, 201, 204
**Iota Draconis b,** 72
**Iridium,** 215, 217
**Iron,** 40, 41, 43, 51, 52, 54, 59, 65, 80, 90, 99, 102, 204
**Ishtar Terra,** 53
**Isomorphism,** 99
**Itokawa,** 145

## J

**Japanese Space Agency (JAXA),** 145
**Jaw,** 106
**Jet Propulsion Laboratory,** 140, 142
**Jupiter,** 36, 47, 56-57, 59, 69, 83, 117, 134, 159, 178, 182, 200-201, 202, 206
**Jupiter Icy Moons Explorer (JUICE),** 203
**Jurassic (period),** 92

## K

**Kennedy Space Centre,** 140, 141
**Kepler (telescope),** 72, 75
**Kepler, Johannes,** 116, 122, 125
**Kepler-10b,** 73
**Kepler-186f,** 73, 74, 75
**Kepler-438b,** 74
**Kepler-442b,** 74
**Kepler-452b,** 73, 74
**Kevlar,** 197, 210
**Kibo (laboratory),** 212
**Kimberella,** 103
**Kittinger, Joseph,** 228
**KOI-4878.01,** 75
**Kuiper belt,** 64, 66, 67, 70, 206
**Kuiper, Gerard,** 66

## L

**La Silla (observatory),** 130
**Laika,** 139, 146
**Landsat (satellite),** 218, 221
**Large Hadron Collider,** 128-129
**Leonov, Aleksei,** 146, 147, 153
**Lippershey, Hans,** 117
**Lithosphere,** 81
**Little Joe (rocket),** 148
**Lovell Jr., James,** 152, 153
**Luna (probe),** 146, 148
   *1,* 146
   *3,* 146
   *9,* 148
   *10,* 148
**Lunar Prospector,** 153
**Lunar Reconnaissance Orbiter (LRO),** 153
**Lunar Rover,** 152-153
**Lyndon B. Johnson Control Center,** 140, 142, 143, 213

## M

**McKay, David,** 111
**Magellan (probe),** 53
**Magmatism,** 94, 97
**Magnetism,** 23, 29, 43, 51, 57, 60, 80, 82-83, 178
**Magnetometer,** 153, 207
**Magnetosphere,** 83, 200
**Manned Manoeuvring Unit (MMU),** 169
**Mariner,** 148, 158, 186, 188, 190
   *3,* 186
   *4,* 148, 159, 186
   *6,* 186
   *7,* 186
   *9,* 188
   *10,* 158
**Marrella,** 105
**Mars,** 36, 37, 47, 50, 54-55, 68, 69, 83, 110, 148, 158, 159, 186-187, 188-189, 190, 191, 192, 193, 196-197, 198-199, 202

**Mars Colonial Transporter,** 198
**Mars Express,** 144, 187
**Mars Odyssey,** 187, 188-189
**Mars One,** 199
**Mars Pathfinder,** 159
**Mars Reconnaissance Orbiter,** 187, 190-191
**Mars Science Laboratory,** 159, 194
**Marshall Space Flight Center,** 140
**Marsnik (probe),** 186
**Matter,** 19, 20, 21, 24, 26, 127, 129
**Mauna Kea (observatory),** 131
**Maven (probe),** 187
**Mawsonite,** 103
**Mayor, Michel,** 72
**McCandless, Bruce,** 169
**Mercury,** 36, 37, 47, 50-51, 83, 158
**Mercury (spaceship),** 147, 148, 149
**Mesosphere,** 81
**Mesozoic,** 92
**Messenger (probe),** 158
**Metamorphism,** 94, 97
**Meteorite,** 68, 69, 90, 92, 110, 139, 187
  *Aerolite,* 68
  *Allan Hills 84001,* 111
  *Murchison,* 111
  *Siderolite,* 68
**Methane,** 59, 60, 61, 62, 63, 65, 100, 186, 205
**Microgravity,** 170, 172, 212-213
**Milky Way,** 16, 17, 26, 28-29, 31, 40, 127, 133
**Minkowski, Hermann,** 127
**Miocene (period),** 93
**Mir (space station),** 144, 145
**Miranda (moon),** 61
**Moon,** 54, 68, 78, 84-85, 86-87, 88, 114, 116, 117, 124, 134,

135, 138, 146, 150-151, 152, 153, 158, 178, 189
**Musk, Elon,** 198

## N

**NASA,** 19, 75, 111, 138, 140-141, 142-143, 144, 146, 148, 151, 152, 153, 154, 172, 174, 175, 177, 179, 180, 182, 187, 188, 190, 194, 196, 197, 198, 199, 200, 202, 204, 206, 223
**NAVSTAR,** 224
**Nebulae,** 38-39, 41, 175
  *Butterfly,* 38
  *Cat's Eye,* 175
  *Helix,* 39
  *Hourglass,* 39
  *Spirograph,* 38
**Neogene,** 93
**Neptune,** 46, 47, 62-63, 65, 66, 83, 158, 159, 178, 179
**Nereid (moon),** 62
**Neutron,** 19, 20, 23, 43, 48
**New Horizons (probe),** 159, 202, 206-207
**Newton, Isaac,** 22, 23, 117, 124-125, 126
**Nickel,** 52, 65, 80
**Nitrogen,** 38, 39, 48, 52, 55, 58, 64, 65, 128, 146, 205, 231
**Nix (moon),** 64
**Nuclear (force),** 23

## O

**Oberon (moon),** 61
**Oberth, Hermann,** 138
**Oligocene (period),** 93
**Olympus (mons),** 55
**Opportunity (rover),** 187, 192-193
**Orbit,** 36, 37, 47, 51, 54, 60, 64, 65, 66, 69, 70, 71, 78, 118, 150, 159, 162, 165, 176, 177, 182, 190, 204, 213, 214, 215, 216, 217, 218, 222, 225
**Ordovician (period),** 91, 106
**Orion (spacecraft),** 197, 198
**Oxygen,** 36, 38, 40, 41, 48, 55, 64, 80, 90, 100, 146, 152, 163, 167, 168, 169, 170, 202, 231

## P

**Paleogene (period),** 92
**Paleozoic,** 91, 106
**Pangea,** 91, 93
**Panspermia,** 110, 111
**Paranal (observatory),** 130-131
**Parsec,** 32
**Pathfinder,** 187
**Pegasi b (51),** 72
**Penzias, Arno,** 25
**Permian (period),** 91, 107
**Phobos (moon),** 54

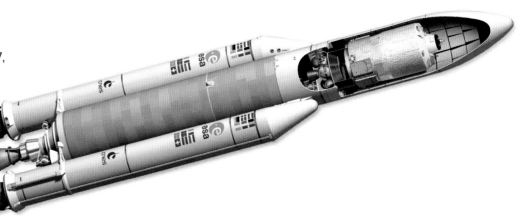

**Phobos (probe),** 187
**Phoenix Mars Lander,** 187
**Photosphere,** 48, 49
**Photon,** 18, 20, 48
**Phylogenetic tree,** 108-109
**Pikaia,** 105
**Pioneer (probe),** 159, 178, 179, 200, 202
  *10,* 159, 178, 179, 200, 202
  *11,* 159, 178, 200, 202
**Planisphere,** 132-133, 134, 135
**Pleistocene (period),** 93
**Pliocene (period),** 93
**Ptolemy, Claudius,** 116, 122
**Pluto,** 64-65, 66, 67, 206, 207
**Polycarbonate,** 210
**Polymorphism,** 99
**Positron,** 19
**Precambrian (period),** 90, 102
**Priapulida,** 104
**Procellarum (Moon ocean),** 84
**Prokaryotes,** 100, 101, 102
**Proterozoic,** 90, 102
**Proton,** 19, 20, 23, 43
**Proxima b,** 72-73
**Pulsar,** 43

# Q

**Quaoar,** 67
**Quark,** 18, 19, 23
**Quasar,** 30, 31
**Queloz, Didier,** 72

# R

**Radiation,** 25, 41, 81, 139, 170, 192, 197
**Radio telescope,** 180, 181
**Redstone (rocket),** 148
**Refraction,** 61
**Relativity (theory),** 22, 43, 126-127
**Rembrandt (crater),** 50
**Ring,** 57, 58, 61, 62
  *Adams,* 62
**Rocket,** 138, 139, 144, 145, 146, 148, 149, 150, 160, 161, 162-163, 164-165, 167, 192, 222
**Rosmos,** 186
**Rotation,** 47, 78, 82
**Rupes Altai (Moon mountain),** 85
**Russian Federal Space Agency,** 145
**Rutan, Burt,** 226

# S

**Sagan, Carl,** 21, 181
**Sagittarius (galaxy),** 29
**Satellite,** 135, 138, 139, 145, 163, 165, 166, 175, 176, 214-215, 216-217, 218-219, 220-221, 222, 223, 224, 225
**Saturn,** 36, 46, 47, 57, 58-59, 61, 66, 83, 159, 178, 204-205
**Saturn V (rocket),** 138, 150, 165
**Schiaparelli (module),** 187
**Schmitt, Harrison,** 152, 153
**Sedna,** 67

**SETI,** 180-181
 **Shepard, Alan,** 147, 148
  **Shuttleworth, Mark,** 226
   **Silurian (period),** 91, 106
   **Silver,** 98
   **Smart (spacecraft),** 153
  **SOHO (observatory),** 175
**Solar System,** 21, 28, 39, 46-47, 50, 54, 56, 57, 61, 64, 66, 68, 70, 71, 73, 74, 110, 124, 152, 158, 159, 178, 181, 202, 206
**Soyuz,** 145, 149, 152
  *1,* 149
**Space Launch System (SLS),** 198, 202
**Space Shuttle,** 142, 143, 158, 161, 166-167, 169, 170, 200
**SpaceShipOne,** 226, 227
**Spectrometer,** 153, 188, 191, 207
**Spicule,** 49
**Spirit (rover),** 187, 192-193
**Spitzer (telescope),** 73, 75, 175
**Sponge,** 104
**Spot,** 218, 219, 220, 221
**Sputnik,** 138, 139, 146, 148, 186
  *1,* 138, 139, 148
  *2,* 139, 146
  *22,* 186
  *24,* 186
**Star,** 16, 20, 24, 26, 27, 28, 29, 30, 32-33, 34-35, 36-37, 38-39, 40-41, 42-43, 72, 73, 75, 114, 115, 118, 119, 126, 127, 132, 134, 135, 154, 175, 207
  *Eta Carinae,* 28, 175
  *Neutron,* 43
  *Red Giant,* 35, 36-37, 43
  *Red Supergiant,* 35, 36, 43
  *Supernova,* 20, 34, 35, 40-41, 175, 179
  *White Dwarf,* 37, 38, 39, 43
**Stratosphere,** 81, 228, 229, 230, 231
**Stromatolite,** 102
**Su Song,** 120
**Sulphur,** 51, 52, 59, 98
**Sun,** 21, 28, 33, 34, 36, 37, 38, 39, 40, 43, 46, 47, 48-49, 50, 51, 52, 53, 54, 58, 60, 62, 65, 69, 71, 72, 75, 78, 82, 83,

86, 87, 88, 89, 115, 116, 117, 122, 124, 125, 126, 127, 158, 170, 171, 178, 182, 183, 191
**Supercluster,** 17
**Surveyor I,** 148
**Swigert, John,** 152

# T

**Tectonic plates,** 93
**Teflon,** 210
**Telescope,** 116, 117, 120, 122, 125, 130, 131, 134, 138, 174-175, 176
**Tereshkova, Valentina,** 146, 147
**Thermosphere,** 81
**Titan (moon),** 58, 110, 158, 159, 204, 205
**Titania (moon),** 61
**Tito, Dennis,** 199, 226
**Tranquility (module),** 173
**TRAPPIST-1,** 73, 75
**TrES-1,** 73
**Triassic (period),** 92
**Tribrachidium,** 103
**Triton (moon),** 62, 179
**Troposphere,** 81, 229, 231
**Tsiolkovsky, Konstantin,** 138

# U

**Ulysses (probe),** 144, 159, 182-183
**Umbriel (moon),** 61
**Universe,** 16-17, 18-19, 20-21, 22-23, 24-25, 30, 115, 117, 122, 124, 127, 128, 129, 174, 180

**Upsilon Andromeda,** 72
**Uranium,** 41
**Uranus,** 46, 47, 60-61, 62, 66, 83, 158, 159, 178

# V

**Valles Marineris,** 55
**Van Allen Belts,** 83, 139, 217
**VASIMR (rocket),** 197
**Venus,** 36, 37, 47, 52-53, 83, 110, 117, 135, 158, 200, 204, 206, 207
**Venus Express,** 144, 158, 207
**Very Large Telescope (VLT),** 130-131
**Viking (probe),** 159, 187, 188
**Voskhod,** 146, 147, 153
 *1,* 147
 *2,* 146, 147, 153
**Vostok,** 146, 147
 *1,* 146-147
 *2,* 147
 *3,* 147
 *4,* 147
 *5,* 147
 *6,* 146, 147

**Voyager (probe),** 61, 62, 159, 178-179, 200, 202, 204
 *1,* 159, 178, 179, 200
 *2,* 61, 62, 159, 178, 179, 200, 204

# W

**White, Edward,** 169
**Wilson, Robert,** 25
**Wimp,** 23
**WMAP,** 19, 154-155
**WOW! signal,** 181

# X

**X-rays,** 30, 31, 41, 42, 43, 176

# Z

**Zarya (module),** 172, 173
**Zvezda (module),** 172, 173
**Zodiac,** 118, 119

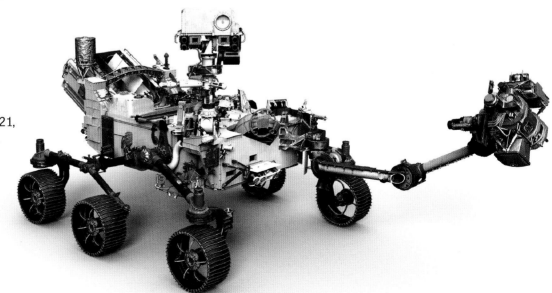